402 "City of Winnipeg" Squadron *History*

402 "City of Winnipeg" Squadron History

Edited by Pat McNorgan

Copyright © 2007 The 402 Squadron Association

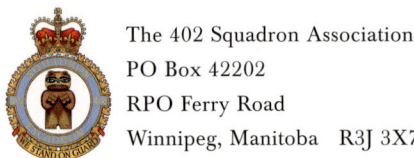

The 402 Squadron Association
PO Box 42202
RPO Ferry Road
Winnipeg, Manitoba   R3J 3X7

Front Cover Photo: Paul Bowen Photography

Book Design and Layout: Adrienne Popke

Substantive Edit and Copy Edit: Evelyn Falk

Library and Archives Canada Cataloguing in Publication

402 "City of Winnipeg" Squadron history / edited by Pat McNorgan.

Includes bibliographical references and index.
Issued by: 402 Squadron Association
ISBN 978-0-9784109-0-2

1. Canada. Royal Canadian Air Force. Squadron, 402--History. 2. Canada. Royal Canadian Air Force--History--World War, 1939-1945. 3. Canada--Armed Forces--History. 4. World War, 1939-1945--Aerial operations, Canadian. I. McNorgan, Pat II. 402 Squadron Association III. Title: City of Winnipeg Squadron history. IV. Title: 402 Squadron history.

UA602.C57F68 2007   358.4'1310971   C2007-980171-4

Printed in Canada

Every effort has been made to ensure the information contained in this book is accurate, however readers are encouraged to send any corrections to the Squadron website www.402squadron.com; they will be placed with a copy of the book in the 402 Archives as a resource to those writing the next history.

Acknowledgements

This book rectifies a major omission from Canadian Forces flying squadron histories. Although 402 Squadron has produced three booklet-form history sketches to mark its 40th, 50th and 70th anniversaries, a formal history had been conspicuously absent. This is a somewhat ironic situation since the Squadron is one of the two oldest active flying units in the Air Force, and has a heritage second to none. As such, this project has been both a duty and a labour of love. Many people have invested substantial amounts of their own time, talent and money in bringing this project to fruition, and so this history, in true Air Force tradition, has been the result of a dedicated team effort.

Four authors were approached to write the Winnipeg Bears' story, and all stepped forward. My thanks to Cpl Jim Bell, Stephen Fochuk, Leo Pettipas (Associate Air Force Historian) and LCol Dean Black for providing the backbone of the book. The editor is deeply indebted to the Office of Air Force Heritage and History and its director, Don Pearsons. A very special thanks to the Air Force Historian, Capt Ben Bond, for his patience, guidance and advice. The father of the project has been LCol (ret'd) Bob Patrick, who as always, pulled things together (and worried so much). Russ Lanoway gave his financial assistance and was one of the driving forces behind this book. I also wish to thank: MCpl Josephine Sallis, for her research, writing and editing of the band chapter; Maj (Ret'd) Michael McNorgan, Dr. Carl Christie (Associate Air Force Historian), Maj Mathias Joost and Dennis Mah at the Central Imaging Library who contributed and identified photographs; Al Chalkley and Norman Malaney, who loaned the outstanding images from their collections; Art "Barny" Barnard and Rick Richards, active voices from the Squadron's war years, for their support and information; 402's Honourary Colonel, MGen (ret'd) Rick Linden; Maureen Walls at the Voxair; and former 402 CO Matt Reid, Hugh Halliday, Bill Davidson, Jerry Vernon and Hugh Braceland.

The following people made financial contributions to both the history book and 75th anniversary weekend. Thanks to: George Richardson; Atlantic Turbines International; Bristol Aerospace; Craddock Farms; Field Aviation Co. Manitoba Hydro; Skyservices; The Winnipeg Foundation; Larry Argue, Wayne Chalaturnyk, Richard Howard, and Cheryl McDonald; Jackie Hall and Barbara, Cameron, Jason & Andrea Bell in memory of Capt. Jack Reeve; Ross Innes in memory of S/L J16925 Bruce Innes DFC; David & Linda Loeb in memory of Capt. Jack Reeve; and Jean Sauder in

ACKNOWLEDGEMENTS

Memory of Mark Sauder; 402 Honourary Colonel (ret'd) Bob James; LCol Rick Witherden; LCol (ret'd) Bob Patrick; LCol Marc Rittinger; F/L (ret'd) John Caron; MWO Louis Viens; WO (ret'd) Tulse Das, WOs Jamie Burke and Dave Sallis; FSgt (ret'd) Laurie McGregor; Sgts Kaye Alex, Blaine Dorie, Olle Fritsch, Gary Lacoursiere, Andi Muralt, Chris Ritchot, Robert Vipond and Troy Zuorro; MCpls Alf Adams, Sean Hubbard, Margaret McKenzie, Josephine Sallis, Stacy Sellsted, Elizabeth Vipond and Ellen Wood; Cpls Johann Ferwerda, Paul Simms, Robert Schwindt and Steve Wong.

To anyone whose name has fallen through the cracks, I offer my sincere apologies.

It is our hope that you will enjoy this squadron history. May it rekindle fond memories and old friendships.

Pat McNorgan,
Editor

Table of Contents

Introduction 1

CHAPTER 1:   Beginning and Pre-war Years 3
                    Corporal Jim Bell

CHAPTER 2:   The Second World War 29
                    Stephen M. Fochuk

                    Hurricane 1940 (poem) 91

402 Squadron War Record 95

CHAPTER 3:   402 Squadron, 1946-1968 97
                    Dr. Leo Pettipas

Flight Sergeant Minto 165

Rank Chart: RCAF - CF Equivalent Ranks 167

CHAPTER 4:   Post-Unification 169
                    Lieutenant-Colonel Dean C. Black

402 Squadron Pipe and Drum Band: The Capture of Hearts 199
    Master Corporal Josephine Sallis

Appendix A: 112 Squadron Officers - Known Commanders and Honours 217

Appendix B: Squadron Commanding Officers 218

Index 219

Introduction

In December of 1932, the Minister of National Defence approved the formation of an air force auxiliary squadron in Winnipeg. The RCAF had been established in 1924 with a provision for a Non-Permanent Active Air Force, but no reserve units were established until 1932. The impetus for authorizing auxiliary squadrons in 1932 was a severe cut in military spending that led to reductions in the strength of the Regular Force. This cut was probably inevitable during the Depression, but the failure of the 1932 Disarmament Conference and the subsequent re-arming of Germany led Major-General Andrew McNaughton, Chief of the General Staff, to recommend the formation of three auxiliary squadrons. Ultimately, three army co-operation units, 10, 11 and 12 Squadrons, were formed in Toronto, Vancouver and Winnipeg respectively. These units were quickly followed by squadrons in Montreal, Hamilton and Regina. These were truly community-based units with few resources, no take-home pay, and little support, but also with tremendous enthusiasm and dedication. By 1939, the Active Auxiliary Air Force, as it was then called, consisted of 12 squadrons and 1000 personnel.

The wisdom of General McNaughton's decision to create a reserve when the military budget was decimated was demonstrated when a mobilization order was issued on August 31, 1939. Within two weeks, almost all eligible members of the Auxiliary had volunteered for service. They made up almost half the strength of the Royal Canadian Air Force. From this small core of people grew over 40 Canadian squadrons serving in Britain, as well as the nucleus of people who organized the massive British Commonwealth Air Training Plan. Members of 112 Squadron (later to become 402 Squadron) were attached to 110 Squadron from Toronto to become part of the first RCAF unit to serve overseas.

This book highlights the fact that change has been a constant feature of the Squadron's history. Members of the Winnipeg Bears went to Europe as part of an army cooperation squadron, flying Lysander aircraft. Shortly after their arrival in England, the Unit became a fighter squadron because there was no need for army support aircraft. Instead, there was a pressing need for fighters and crews to resist the German air offensive. Thus, instead of going directly into combat, the members of 402 had to relocate and learn to fly and maintain Hurricane fighters. Throughout the war, the Squadron was constantly moving, and converting to newer and better aircraft. As the Allies obtained air superiority, the mission changed from defensive to offensive operations, and 402 began escorting bomber missions over Germany. In September of 1944, the Squadron moved to Holland, where they continued to fight the Luftwaffe and conducted large numbers of armed reconnaissance missions.

Constant change also characterized the postwar period. The role, structure and equipment of 402 changed in response to changes in the nature of the threat and the strength of the Canadian economy. Immediately after the war, the Auxiliary formed the backbone of the RCAF, and 402 remained in a front-line role until 1956, when the switch from Mustangs to Expeditors marked the beginning of a long period when 402 and the rest of the air force reserves marked time while the military relied upon the regular force to meet the challenges of the Cold War. During my own time with 402 during the 1970s and 1980s, a very dedicated and highly professional group of men and women worked with badly outdated equipment and no clear role, while developing the standards that are reflected in the operation of 402 Squadron today. Their professionalism had an even more important impact. The success of the Air Reserves during the 1980s and 1990s also set the stage for today's total force military where the reserves are closely integrated with the regular force in all three services. Today, the Canadian Forces depends upon reservists for a broad range of ongoing activities, including its operations in Afghanistan.

The changes have meant that people from different eras in 402's history have had very different experiences with the Squadron. The pilots and crews operating Mark IX Spitfires out of RAF Station Kenley were involved in very different kinds of activities than the 402 members that flew Expeditors in the '60s, Otters in the '70s, Dakotas in the '80s and the Dash-8 navigation trainers in the new millennium. The large complex wing structure of the 1950s, made up almost entirely of part-time personnel, was very different from today's total force unit, which has a high proportion of full-time regular force members. However, despite all the changes, there has been a common bond of discipline and professionalism, and a passion for excellence that links the different generations of those who have served with 402.

The authors have performed a valuable service by chronicling much of the history of 402. This volume will help to strengthen the proud traditions of the Squadron as it evolves in the future.

Major-General (retired)
Eric (Rick) W. Linden, CD

Honourary Colonel
402 Squadron

THE DE HAVILLAND GIPSY MOTH

J. McNulty Collection

Beginning and Pre-war Years

Corporal Jim Bell

The officers and other ranks of 112 (Army Co-operation) Squadron, Active Auxiliary Air Force, assembled in Winnipeg's west end Minto Armouries on Sunday, 3 September 1939. Earlier that day, Great Britain and France had declared war on Nazi Germany. During a private meeting with the commanding officer, Squadron Leader (S/L) H.P. Crabb, all the officers had agreed to leave their civilian jobs to go on full-time service. Meanwhile, the other ranks fell in on parade to hear the Squadron's permanent force adjutant, Flight Lieutenant (F/L) D.M. Edwards, announce that the unit was now on active service designation.

"Anyone who is not available for active service, take one pace forward." Out of 100 personnel, only one man moved; 112 Squadron was going to war.

When the Royal Canadian Air Force (RCAF) formed on 1 April 1924, it included provision for a Non-Permanent Active Air Force (NPAAF). Officers and airmen were taken on auxiliary strength, despite the fact that no units yet existed. Individually, these personnel were used primarily on the Civil Government Air Operations that occupied so much of the Air Force's time during the late 1920s and early 1930s.

CHAPTER 1

Lobbying the federal government for units soon began. As early as 1925, the Air Force Club of British Columbia urged Parliament to consider the formation of a "Reserve Air Arm" to "guard the coast of British Columbia which lies nearest the most probable zone of warfare."[1]

These lobbyists often understood air power issues from personal experience. For example, when 11 Squadron stood up on 5 October 1932, Major A.D. Bell-Irving, a prominent First World War pilot and president of the Air Force Club, became its first commanding officer.

More people than the arguably biased members of The Air Force Club wanted auxiliary squadrons. Well-known pilots, financiers, and military men came together in 1927 to form the Aviation League of Canada. The League had six aims, including the "provision of an Air Force adequate to meet Defence requirements."[2] The Aviation League included on its board such people as R.H. Mulock, the Winnipegger who commanded the RAF's Independent Force at the time of the Armistice; the famous air ace Billy Bishop, one of the best known Canadians of the day; and James A. Richardson, who in 1926 formed Western Canadian Airways, perhaps the

R.H. MULOCK

BILLY BISHOP

country's first successful commercial airline. Officials of the league maintained an office in Ottawa, keeping in close contact with the Department of National Defence. Its mailing list included the prime minister, cabinet members, and provincial premiers. Besides the League, individual private citizens who felt that, "Canada in general and their city in particular should have an Air Force equivalent to the Army and Navy reserve units,"[3] provided another source of pressure on the government.

These efforts went for naught until the Great Depression struck. In 1932, the government slashed the RCAF budget, resulting in substantially reduced operations. Almost 200 officers and airmen left the non-permanent list. To mitigate the effects of the cut, the RCAF began planning the formation of auxiliary squadrons. It promulgated regulations for non-permanent squadrons, and raised the non-permanent establishment from 85 officers, and 130 airmen, to 128 officers, and 624 airmen – almost the same strength of the permanent RCAF.

In September 1932, the Chief of the General Staff (CGS), Major-General Andrew McNaughton, recommended to the Minister of National Defence that the NPAAF encompass one wing and three squadrons. The Governor-in-Council approved this recommendation on 5 October 1932, with all new units designated for army co-operation duties. McNaughton's recommended wing did not materialise, but 10 Squadron was earmarked for Toronto, with numbers 11 and 12 "to be localized hereafter."[4]

In November or December 1932, the Winnipeg Board of Trade sent a letter to the Senior Air Officer, Group Captain (G/C) J.L. Gordon, requesting an auxiliary squadron for their city.

The letter pointed out Winnipeg's advantages: an aircraft industry, an airport, qualified and experienced manpower, the Headquarters of Military District 10 and an RCAF depot workshop, which could provide assistance if necessary.

On 14 December 1932, Gordon wrote to the CGS echoing these points, adding "An application has been received from Vancouver for the localization there of one additional squadron, and this request from Winnipeg would, if approved, together with the request already submitted in the case of Vancouver, give effect to the localization of each of the three squadrons already approved by Council."[5] It appears from this communication that no other city applied to receive a squadron, and that the Department of National Defence made no effort to identify potential sites. How the Board of Trade found out about the formation of auxiliary squadrons is unknown. Presumably, James A. Richardson or John Sully, another Aviation League member, heard about it and convinced the Board to write the letter.

The Minister approved Gordon's recommendation on 16 December, with the proviso that initial organisation of the three squadrons incur no additional expense to the public, save for specially approved small sums. On 4 January 1933, the Adjutant-General wrote to the 10 District Officer Commanding (DOC) advising him about the formation of an air force auxiliary squadron in Winnipeg.

A second letter, dated 23 January, relayed the details omitted in the first message, giving the DOC responsibility for administration, discipline, and finding suitable accommodation for the Squadron. Most importantly, he had to recommend an officer for appointment as 12 Squadron's first commanding officer.

John Alfred Sully, Air Force Cross (AFC) may have been the obvious choice. During the First World War, he had served in the Canadian Expeditionary Force, later transferring to

S/L SULLY - 12 SQUADRON'S FIRST COMMANDING OFFICER

the Royal Flying Corps as an observer. He distinguished himself by shooting down one enemy aircraft and damaging two more, in a single action. Before the war ended, he had won his pilot's wings. In the 1920s, he carried on a large farming operation, while simultaneously running an insurance company. Moving to Winnipeg, he joined the London Life Insurance Company, becoming a district manager. He involved himself with the organization of the Winnipeg Flying Club and became its first president. Further, he helped in the formation of the Canadian Flying Clubs Association and served as vice-president of the Aviation League.

With the rank of Squadron Leader (S/L), John Sully became the initial member of 12 Squadron on 1 March 1933. Upon Sully's appointment, the real work of organising a flying squadron began with the recruitment of officers.

By autumn, F/L H.P. Crabb, a wealthy businessman and member of the local flying club with First World War experience, had joined 12 Squadron. Indeed, Crabb's credentials made him the prototypical auxiliary officer. By the end of the year, Squadron strength stood at four: Sully, Crabb, and F/L's Huggard and H. Little.

A great deal of work went on behind the scenes. Officer enrolments, preparation of training plans, lessons and other support functions kept Sully and his growing staff quite busy. Minto Armouries set aside four rooms for the Squadron and the airmen practiced drill when the parade square was available. F/L K.M. Guthrie,[6] the station adjutant, provided administrative assistance for Sully to the detriment of his regular duties. Nevertheless, the Air Force denied an October request for an administrative clerk, due to lack of manpower at the station.

Unofficial training probably began in September 1933. A lack of training aids or qualified instructors limited the training to drill and service subjects. There were plenty of trainees though. Despite a roster of only four officers, a typical 12 Squadron parade night found dozens of young men at Minto Armouries, all in civilian clothes, all unpaid, and none with any idea when they might become members of the RCAF.

In January 1934, the Squadron began recruiting. For the sixty positions authorized, the air force provided only fifty uniforms, and no compensation (it all went to the Squadron Fund). With the exception of summer camp, members were expected to sign their pay over to the Squadron. The pay sign-over was not negotiable. New officers also had to purchase their uniforms within three months of enrolment. General List Officers (flying) were required to have previously earned their wings, or hold a civilian pilot's licence.

The waiting list to get in was long, and it helped if the prospective airman knew someone in the unit. Art Johnson's boss worked with Pilot Officer (P/O) Patriarche, an officer who joined early in 1934. A young mechanic named Frank Klaponski worked at a garage where F/L Huggard had his automobile serviced. Klaponski gave it extra attention

SQUADRON AIRMEN TAKE A BREAK, CIRCA 1938. NOTE THE HIGH BUTTON TUNICS UNDONE AT THE NECK FOR COMFORT.

before delivering the car to Huggard. Impressed with the young man's efforts, Huggard asked Klaponski if he knew about the auxiliary squadron. Frank replied, "Yes, I've been on the waiting list for months." Huggard told him to come down to the armouries, and Klaponski was in.

Getting to the top of the waiting list did not guarantee a place in the Squadron. Recruits had to pass a two- to three-month probation period during which they were not in the Squadron but in the "awkward squad." These members were immediately dismissed if they missed a parade night for any reason.

Training consisted mostly of general service knowledge subjects and drill, given by the Permanent Force "discip". Trades training began after the probationary period, however, most members came to 12 Squadron with a solid work ethic.

RIFLE DRILL AT CAMP SHILO 1936

A letter from the DOC to the Secretary of the Department praised the members of the new squadron:

> The enthusiasm and willingness to work shown by both the officers and the airmen is remarkable. All officers, whether actually appointed to a commission, or only recommended for an appointment have regularly attended all lectures and drill since early in January 1934, with the clear understanding that individual appointments may not be authorised.

A lucky few, such as Andrew Craig, avoided probation. He tried to enroll in the permanent RCAF in 1935, but since there were no vacancies, the recruiting officer told him to join the auxiliary squadron. He enrolled directly as an aircraftman second class (AC2) and stayed with the unit until a vacancy occurred in the Permanent Force in December 1937.

With its growing establishment, 12 Squadron needed help to carry out training. In January 1934, RCAF Station Winnipeg tasked a clerk and a drill instructor to assist one evening a week. Before long however, it was obvious that this was not enough and, in October 1934, 12 Detachment formed to support 12 Squadron. The detachment consisted of one officer, who served as adjutant, and a number of other ranks.

The detachment's first commander was Flying Officer (F/O) E.A. McNab, who went on to lead the RCAF's only Battle of Britain squadron (No.1 Fighter Squadron) and received Canada's first Distinguished Flying Cross (DFC) of the war.

The second commander was former Siskin pilot F/O F.M. Gobeil. When the war began, he commanded 242 (Canadian) Squadron, Royal Air Force (RAF) and scored the RCAF's first aerial victory when he shot down a German Me110 over Belgium on 25 May 1940. Both McNab and Gobeil served a year in 12 Detachment.

F/O D.M. Edwards, who stayed with the Squadron until the war began, replaced Gobeil. The other ranks instructed or supported the auxiliary personnel.

The quickly established routine saw all training taking place between 2000 and 2215 hours. At first, trades instruction took place on Tuesday evenings, however as the Squadron expanded it became necessary to break training down by flights. From September 1934, "A" Flight paraded on Mondays, "B" Flight on Tuesdays, and "C" Flight on Wednesdays. Thursday nights, the Squadron held parades.

"A" FLIGHT DURING SUMMER CAMP IN SHILO

Each flight followed an identical schedule. For example, on the first three days of October 1934, A, B, and C Flights trained in wood construction for riggers, (air frame mechanics) bench work for fitters, (aero engine mechanics) accounting for storesmen, elementary electricity and wireless telephony practice for signalers, and elementary machine-gun training for armament artificers.

PRE-ENLISTMENT TRAINING ON HENRY AVENUE IN WINNIPEG - CHIEF INSTRUCTOR CONNIE JOHANNASON WORKS ALONG WITH ALF WAITS

Thursdays, the entire Squadron met for an hour of drill followed by various classroom subjects, which included air force organisation, history, and technical movies. After these sessions were finished, many members retired to their various messes. This social aspect of Squadron life was instrumental in keeping personnel committed.

THE DE HAVILLAND GIPSY MOTH

In September 1934, the Squadron received four unassembled de Havilland DH-60GM Moths. The Gipsy Moth, a little two-seat biplane designed in 1925, became the unit's first aircraft. It could attain a maximum speed of 105 mph at sea level, a cruising speed of 85 mph and a service ceiling of 18,000 feet. Safe and easy to fly, the Gipsy Moth could be built and repaired with ordinary woodworking tools.

The new aircraft arrived in Winnipeg by train from Camp Borden. Trucks delivered them to Stevenson Field for assembly over the next several weeks. With the arrival of the aircraft, the Squadron took over the RCAF hangar, which proved barely large enough to hold the four Moths with their wings folded. From October onward, airmen could spend their Saturdays and Sundays at the aerodrome.

AN ORIGINAL NIGHT LANDING LIGHT LOCATED AT STEVENSON FIELD. NOTE THE FLYING CLUB ROOMS BEING BUILT IN THE BACKGROUND, CIRCA 1935-6

Ernie Moncrieff was typical of the new pilots joining 12 Squadron. He joined in the autumn of 1934 as a Pilot Officer, and remained until May 1940. Although Moncrieff had no prior military experience, he did hold a pilot's licence. His first flight, with F/O McNab, took place on Sunday, 4 November, in Moth 161. Over the next ten months, Moncrieff accumulated fourteen hours, forty-five minutes solo, and four hours ten minutes dual. He flew mostly on Sundays, and completed his wings testing during four flights in July and August 1935. The test included short and long (one and a half hours duration) cross-country flights, along with forced landings. Moncrieff went on to command 430 Squadron in 1943. As G/C Moncrieff, AFC, he commanded 39, (Reconnaissance) Wing until February 1945.

The Moths proved adequate for the first two years of flying operations. Pilots flew circuits or cross-country flights with Lac du Bonnet, sixty miles away, becoming a favourite destination. Typically, one or two pilots would fly to the RCAF station have lunch, and then return. Training in army

CAMP SHILO IN 1936 –
LEFT TO RIGHT HINOM, J. MEEK, LINK, PROUX

co-operation work did not take place during the first years. The Gipsy Moth could not perform army co-operation work such as bombing, photographic, gunnery, wireless, night flying, navigation, and instrument training.

THE DE HAVILLAND GIPSY MOTH

From October 1934 to March 1935, 12 Squadron flew 119.25 hours. In Toronto, 10 Squadron flew 119.1 hours and Vancouver's 11 Squadron flew 248.3 hours. Varying weather no doubt contributed to the difference in flying hours between Vancouver and Winnipeg. In the fiscal year 1935-36, the auxiliary squadrons' authorized flying time increased to 800 flying hours each.

READY TO GO FLYING – 1936
GIPSY NUMBER 60 WITH L. WHITMAN AND GERRY FRASER

Winter in Winnipeg presented many challenges to flying. According to aircraftman Bennie Proulx, the hangar was not heated. When flying was finished for the day:

CAMP SHILO FROM THE AIR DURING 112 SQUADRON'S SUMMER CAMP 1938

The oil was drained from the aircraft, in a five gallon container, marked with the serial number of the aircraft, and placed close to the Quebec stove. And the following day, or whenever you wanted to use the aircraft, you'd select the aircraft you were going to fly, make sure it had been inspected for flight, and park the aircraft as close to the hangar door as possible. When everyone was ready, you'd take this oil that you had previously put close to the furnace to warm up to room temperature if possible, and pour it into the aircraft oil tank. Then you'd push the aircraft outside, close the hangar door, then the prop could be swung. You'd spread ashes on the ground so the person swinging the prop wouldn't slip and fall and get injured.

Pilots wore a chamois facemask, which, except for eye slits, covered the entire face as protection against the cold and windburn of the slipstream. The Gipsy Moths were fitted with skis for winter flying, and taxiing the aircraft required an unusual technique. Aircraftman Vince Marrin explained:

> They couldn't turn. We used to go out and pick a crust of snow and build kind of a wall against the wind, and bury ourselves in the snow. The pilots knew where we were, so they'd land, and we'd run and grab the hand hold on the wing. Then they'd gun the motor, and they could turn.

Squadron strength increased to eleven officers and ninety-six airmen, with an establishment of sixteen officers and 113 airmen by early 1935. Air Force Headquarters also decided that since training was sufficiently advanced, the Squadron could attend its first summer camp. On Friday, 28 June 1935, ten officers and about sixty other ranks proceeded to Camp Shilo, located one hundred miles west of Winnipeg.

PEELIN' SPUDS - KITCHEN DUTY AT CAMP SHILO 1936

Although P/O Moncrieff's logbook shows only two days of camp flying, on one of those days, six flights (mostly in Moth 161, totalling two hours and forty-five minutes) were entered. The first flight started at 0805 hours with his last flight ending at 1935 hours. The next day he made a single flight in Moth 105, lasting one hour fifteen minutes.

The camp days were long with flying beginning at 0600 hours, and concluding at 2000 hours. Each flight's duties included spending one day on the flight line, one day on camp work, and one day training. The flight line work for the most part consisted of fuelling and inspecting the aircraft.

Four officers and four airmen flew the aircraft to the camp, while the remainder travelled by rail. Once at Camp Shilo, Bell tents were erected to accommodate personnel, the aircraft were picketed, and camp routine began.

Officers' training consisted of lectures on airmanship, piloting, map reading and musketry, along with attendance at artillery and cavalry demonstrations. A "special A. C. (army co-operation) aeroplane" performed an army co-operation demonstration, and of course there was plenty of flying.

CAMP SHILO AIRMEN IN 1937

F/O PENTLAND RETURNING FROM A FLIGHT AS HIS MOTH IS BEING REFUELLED AT CAMP SHILO 1936

MACHINE-GUN PRACTICE AT SHILO

An airman's daily routine included kitchen work, camp clean-up, and sundry other jobs. On training days, the airmen learned marksmanship on the No.1 Lee-Enfield rifle. Signallers received training in their specialty, and armament artificers trained on Vickers and Lewis machine guns. Personnel designated as air gunners also trained on these machine-guns. The routine of summer camp was broken by one sports afternoon.

WIRING TRAINING AT CAMP SHILO – EXPLODING PUFF TARGETS FOR ARMY CO-OP TRAINING

In December 1935, James A. Richardson became the Honourary Wing Commander of 12 Squadron. Richardson, being a Winnipeg financier, gave substantial monetary and moral support to aviation associations such as the Aviation League. Richardson's role in the formation and early operation of 12 Squadron cannot be determined because unfortunately the documents have not survived. Once appointed Honourary Wing Commander, Richardson became well-known around the Squadron. He attended each annual inspection, most flying competitions and even an officer's wedding. His support was not limited to just encouragement. At each annual parade, resplendent in top hat and tails, he presented a cheque of $500 or $1,000 to the commanding officer. His total contribution is unknown, but it must have been significant considering the budget for the Squadron, including pay, for the year 1937-38 totalled only $8,811, along with an allotment of $7,207 for the '37 summer camp.

112 SQUADRON TAKING PART IN THE ARMISTICE DAY PARADE 1935

There was good co-operation between Canadian Airways (formerly Western Canadian Airways) and the RCAF in Winnipeg. The RCAF frequently used the Airways' fuel caches in northern Manitoba, and when the Squadron outgrew its hangar in 1939, Canadian Airways gave theirs up for the unit. No doubt Richardson's influence played a major role in the relationship. Sadly, Richardson passed away on 26 June 1939, at the age of fifty-three.

Auxiliary service had its advantages with adventure, friendship, and training, but the disadvantages were substantial as well. The Squadron made heavy demands on their members' time. An airman could spend two evenings plus a weekend day at the unit, thirty-five weeks per year without compensation.[7] Their pay went to the Squadron Fund. The commanding officer of 11 Squadron estimated that an airman who attended regularly should receive 52 days' pay. Six days' pay was provided for airmen, while officers were funded for nine.

In spite of these drawbacks, 12 Squadron had no problem attracting recruits. Retaining them however, proved more difficult, and was exacerbated by the transient nature of employment in the aviation industry, and the lack of modern equipment. A report by higher command on both 11 and 12 Squadrons' signals organisations stated that, "difficulty is found in retaining highly qualified amateur operators in the squadron due to the fact that service equipment is not, and cannot be, so modern as the type of equipment as this type of man is in the habit of using."

The wireless operators of 12 Squadron found a solution by building their own radios to use in the aircraft. Several years later, auxiliary service retention remained a problem. W/C A.H. Wilson, the commanding officer of 111 Squadron

WIRELESS TRAINING BEING CARRIED OUT AT CAMP SHILO. DURING PUFF TARGET EXERCISE THE WIRELESS SET RECEIVER OBSERVATION OF FIRE WAS SENT IN MORSE CODE BY THE PILOT OF THE AIRCRAFT. PUFFS OF WHITE POWDER SMOKE CAN BE SET OFF ON THE GROUND AT VARIOUS POINTS, THEN THE PILOT REPORTS THE FALL OF THE SHOT IN RELATION TO THE TARGET.

(formerly 11 Squadron) wrote in 1938 "assuming that the purpose of the Non-Permanent Active Air Force is to create a pool of trained personnel, it is felt that this unit is not accomplishing this purpose."[8] Referring to all auxiliary squadrons, Air Marshal (A/M) Croil, the Senior Air Officer in 1938, wrote "there is an annual turnover in airmen personnel amounting to approximately thirty per cent which does not permit the technical airmen of the squadron being trained to the standard required."[9] Of one hundred fifty airmen enrolled by 111 Squadron up to May 1936, only twenty-three were still in the squadron in July 1938.[10]

A/M GEORGE MITCHELL CROIL, CBE

S/L Sully's 1935 solution to the turnover problem involved training non-aircraft trade recruits such as cooks, general duties personnel, or batmen, in aircraft trades. Thus, when an airman left the Squadron, a trained replacement could take his place. The downside of this meant the unit became incapable of operating on its own. At its first summer camp in 1935, the Squadron hired four civilian cooks. Although told quite firmly to follow correct procedures, civilian cooks continued to be hired every year for summer camp. This practice caused considerable confusion when the war broke out in 1939, because so many airmen were untrained in their designated trades.

LIFE IN THE BARRACKS

Summer camp, although voluntary, also allowed the only opportunity for auxiliary personnel to receive pay for their efforts. Nevertheless, attendance was often disappointing. In 1936, less than half of the Squadron's airmen turned out. Air A/M Croil noted, "the principal reasons appear to be inability to get time off, or lack of notice."

CAMP SHILO PARADE IN 1937

Attendance did not improve. In 1937, only 49 of the Squadron's 149 other ranks attended the camp. The problem seemed to be resolved in 1938, when attendance at camp jumped to over 100 airmen.

By October, the establishment allotted the unit 16 officers and 149 airmen. Since most of the Squadron's pilots were qualified to wings standard, service training now began in earnest. This training involved instrument flying along with army co-operation work. RCAF Station Winnipeg operated a Fleet 7 Fawn for instrument training. The Fawn was a two-seat biplane primary trainer with excellent flying characteristics. Since the Moth was totally unsuitable for army co-operation work, the unit attempted to have the Fawn transferred to them. Although this request was denied, it did result in the RCAF shipping Fawn 205, from Camp Borden for temporary use. The Squadron used this aircraft from June 1936 to July 1937.

THE FLEET FAWN

SERGEANTS PICKERING AND BROWN AT CAMP SHILO IN 1936

Army co-operation training comprised of classroom lectures in air photography, wireless telegraphy, artillery reconnaissance, and other subjects.

Flying training included instrument work, air reconnaissance, cross-country navigation, and communicating with ground units.

At summer camp, most officers had the opportunity to plot artillery shoots, and take oblique air photos. Air firing of machine guns did not take place due to lack of a suitable range, but the unit did make progress towards becoming a bona fide army co-operation squadron.

PUFF TARGET PRACTICE AT CAMP SHILO, 21 JUNE 1936

PRACTICE BOMBING PHOTOGRAPHY AT CAMP SHILO

The Squadron received a puff target trainer for the army co-operation work-up, which it used to simulate artillery battery shoots. From its trailer, small charges were reeled out, placed in a pattern, and fired simultaneously. The charges resulted in several puffs of smoke, which the pilot could then plot. The puff target trainer saw extensive use during the pre-war years.

Trades training carried on with a traveling trade board composed of permanent and auxiliary personnel dropping in on the Squadron to test tradesmen.

The board visited all auxiliary squadrons ensuring that training standards achieved a service wide consistency. Many tradesmen were now qualified in group B or C, which required advanced training and carried higher rates of pay. Group C tradesmen could sign for maintenance on an aircraft, making the unit less reliant on its permanent force personnel.

At this time the Squadron also began training air gunners. The Princess Patricia's Canadian Light Infantry, based in Winnipeg, provided a non-commissioned officer (NCO) for several months in the spring to give instruction on machine guns.

SERGEANT HARDY WORKING ON AN AVRO TUTOR'S MACHINE-GUN

THE FIRST RCAF FIRE TRUCK IN WINNIPEG, PARKED IN FRONT OF THE MINTO ARMOURIES

This training was limited to theory of operation and weapons handling. Actual air firing of the weapon came later.

From 1936 until 1939, the unit began to make significant progress towards becoming an effective army co-operation squadron. Starting in 1935, the Squadron attempted to send all its officers on a two-week army co-operation course.

The Permanent Force version of the course lasted six weeks, and a few auxiliary officers may have attended it. Armament, wireless or photographic courses were available for officers and other ranks as well.

The 1936 summer camp marked the beginning of routine artillery spotting flights. The success of these flights led to joint exercises with the artillery in Shilo. A typical sortie consisted of two aircraft, each with a pilot and observer, flying out of Winnipeg in the morning. One aircraft would land at Shilo, so the pilot could co-ordinate with the artillery, while the other aircraft did the spotting. Radio messages had to be passed to the pilot on the ground, because aircraft and army radios were incompatible. Message drops were also used to pass information. Once the shoot finished, the observing aircraft landed. After lunch, the roles were reversed. At the end of the day's firing, the aircraft flew back to Winnipeg, usually arriving before 1600 hours.

During the 1938 summer camp, the Squadron co-operated with several army units. During one such exercise, an

G/C G.O. JOHNSON, MC, OFFICER COMMANDING WESTERN AIR COMMAND, INSPECTS 112 SQUADRON AT CAMP SHILO IN THE LATE 1930s. WITH HIM IS F/L E.H. EVANS OF THE PERMANENT FORCE DETACHMENT

THE PROPER FORM FOR DROPPING A FLOUR BOMB BEING DISPLAYED FROM THE REAR SEAT OF A 112 SQUADRON AVRO TUTOR

army service corps unit hid under camouflage near some buildings, and an aircraft unsuccessfully attempted to find them. The next day, vehicles from the same unit tried to avoid a flour-bombing raid, but this time had to concede that aircraft could do considerable damage to vehicles on the move. The camp also featured aerial photography, spotting, and message dropping sorties with a cavalry brigade.

"Much useful information was obtained by both the Cavalry Brigade and the Squadron and particularly the pilots since they learned what mounted troops moving over open country looked like."[11] Still more exercises involved artillery shoots, and reconnaissance of a road move by the 2nd Armoured Car Regiment.

Puff target training took place on the prairie south of Winnipeg, either in conjunction with the permanent or non-permanent militia, or as a squadron operation. The Squadron's photo section also kept busy with pilots practicing vertical and oblique photo techniques. These summer camps provided the Squadron's pilots the opportunity to fly realistic reconnaissance sorties.

Summer camps also gave all ranks a chance to either hone their shooting skills or fire a weapon for the first time. Officers fired pistols, or machine guns on ground mounts, while other ranks had time on the Squadron's SMLE Number Ones (.303 calibre, Short, Magazine, Lee-Enfield, Mark 1), the First World War's standard infantry rifle.

Qualified pilots attending the camps received training in instrument flying, aerobatics, cross-country, and formation flying.

FRED STEVENSON AFTER WHOM WINNIPEG'S STEVENSON FIELD WAS NAMED

The new pilots, still working towards their wings, worked on circuits at Stevenson Field. All this activity created a busy airfield, and by 1938 it had become a problem.

CAMP SHILO DURING THE SUMMER OF 1939

Winnipeg, in addition to being a training centre for Trans-Canada Airlines, became the destination for numerous mail flights, and there was more than one near miss. The field was close to the city of St. James, where complaints about low flying were common. The RCAF search for a new airfield ended with the outbreak of the war.

SQUADRON CAMARADERIE

12 Squadron did not have a problem with morale. Despite the demands the unit placed on its members, officers and other ranks enjoyed their time in the unit. The officers benefited from free flying time and interesting tasks, while the airmen received useful training and developed a spirit of camaraderie. In fact, many airmen joined simply because their friends were already in.

As the 1936-37 training year progressed, the Squadron held several social functions typical of the time. An "all units" social and dance, held on 5 December, inaugurated the

THE 112 SQUADRON BAND

Holiday Season. An officer's mess dinner on the 11th, added to the festivities, followed in short order by a Squadron smoker (a social gathering with food and drink) on the 17th. Two days later, the children's Christmas party provided the Yuletide finale. In January, personnel and their "wives and lady friends" attended a toboggan party and dance. Two weeks later, there was a scavenger hunt, with refreshments afterwards.

On 19 March, the other ranks held a dance, with officer's mess dinners taking place on 25 March and 29 May. The May dinner was an elaborate affair at the Royal Alexandra Hotel and featured the mayor of Winnipeg as the guest of honour.

The airmen held another smoker on 4 June. All of these activities are recorded in the squadron diary.

Each year since 1936, the pilots flew in exciting one-day air shows. Wives and lady friends cheered, while James Richardson and guests watched a flying demonstration followed by a forced landing competition. This involved the pilots approaching the field, cutting their engines, and attempting to roll to a stop on a marked spot.

The bond between the Squadron and the community continued to grow. The unit supported a rifle association and hockey team. The rifle association competed with other military squads, while the hockey team faced off against the police and fire departments. On occasion, the manager of a theatre invited the entire Squadron out to see a movie.

On 17 September 1936, the Squadron marched through the streets on their way to the Capitol Theatre to see "China

112 SQUADRON ENJOYING A MOVIE NIGHT IN WINNIPEG

Clipper", a film starring Humphrey Bogart and Pat O'Brian about the development of Trans-Pacific air routes.

RCAF Headquarters recognised that the Moth aircraft were unsatisfactory for any role other than elementary training. Although plans had existed for some time to equip the auxiliary squadrons with proper army co-operation aircraft, none were available until August 1937. Two Avro 621 Tutor advanced trainers arrived in pieces by boxcar. These were used for service training. Compared to the Moth, the Tutor was larger, more powerful, and more complex. Pilots found the Tutor, another two seat biplane, easy to fly and fully aerobatic. A Mongoose engine (without a cowl) powered the aircraft, enabling it to reach a maximum speed of 120 mph at sea level and a service ceiling of 16,000 feet.

THE AVRO TUTOR

assembly of aircraft due to absolute inexperience of most of the airmen." Despite the diarist's impatience, Avro 224 took to the air within two weeks, and Avro 184 followed with its first flight in early October. Both aircraft were used for service training.

The Squadron received its first true army co-operation aircraft with the arrival of the Avro 626, tail number 226, in March 1938. Avro re-designed the Model 621 Tutor making

THE AVRO 626

Because of its complexity, putting the aircraft back together did not go smoothly. On 25 August, the obviously frustrated squadron diarist noted that, "progress extremely slow on

TUTOR AIRCRAFT LINE-UP. NOTE THE MESSAGE PICK-UP HOOK ATTACHED TO THE UNDERCARRIAGE

THE AVRO TUTOR

it suitable not only for flying training, but also for bombing, photographic, gunnery, navigation, wireless, night flying, and instrument training.

In June 1938, Tiger Moth 248 arrived for use as an elementary trainer. Avro 626's 266 and 267 were delivered in June 1939, with each new aircraft replacing an older one. When the final Avro 626 arrived, the last Gipsy Moth, tail number 74, was retired.

THE DE HAVILLAND TIGER MOTH

Throughout the pre-war period, the Squadron operated a total of four or five aircraft at any given time. The number of authorized flying hours increased as the war drew closer. From 600 hours in 1936-37, it was increased to 800 in 1937-38, and 1,000 hours in 1938-39 and in 1939-40.

After four and a half years as commanding officer, S/L Sully left the Squadron, relinquishing command to S/L Crabb on 1 October 1937. Sully went on to become the first commander of 102 (Army Co-operation) Wing in Montreal. At the time, he was also the Montreal district manager for London Life.

During the Second World War, Sully organised RCAF recruiting stations, commanded RCAF Station Trenton, and became the Air Member for Personnel on the Air Council. As a member of the Air Council, he worked alongside famous RCAF officers like Breadner, Croil, Curtis,[12] Leckie, and Stedman. After being appointed a Companion of the Most Honourable Order of the Bath in 1944, Sully finally retired from the RCAF in April 1945 as an Air Vice Marshal (A/V/M).

Under the original numbering system allotted to the NPAAF, the size of the Permanent Active Air Force was limited to nine squadrons. Since expansion had been forecast beyond this number, and to avoid confusion in numerical designation, all NPAAF squadrons were renumbered by the addition of 100. Thus on 15 November 1937, 12 Squadron became 112 Squadron. Just over a year later, the auxiliary traded one awkward name for another. On 1 December 1938, the Non-Permanent Active Air Force became the Auxiliary Active Air Force.

With the Second World War on the horizon, changes were in store for the Squadron. Tensions rose through August 1939 as Germany tried to provoke Poland into a fight.

When the Germans and Soviets signed their non-aggression pact on 23 August, war seemed inevitable. As the situation deteriorated, Great Britain, France, and Canada moved their troops to a war footing, and 112 Detachment cancelled all leave on 25 August. Three days later, the Winnipeg Light Infantry, a non-permanent militia unit, was called out to guard vital points in the city, including the Squadron's hangar.

Germany invaded Poland on 1 September 1939, and two days later Great Britain declared war. While Canada did not declare war until 10 September, her military forces were ordered onto active service on 3 September.

For the Squadron, the war brought more activity and greater responsibility for the non-permanent personnel. Air Commodore (A/C) Breadner wrote to the Air Officer Commanding Western Air Command on 15 September 1939, instructing him, "Until such time as service aircraft become available for issue to AAAF squadrons, such units are to carry on as heretofore, but on a more intensive basis in keeping with full-time employment."[13] The Squadron's medical staff, two doctors from the Royal Canadian Army Medical Corps (RCAMC) kept busy completing medical inspections on all personnel. A few individuals found unfit were released. The detachment signed over all of the squadron's kit to non-permanent personnel.

Postings also began with F/O Moncrieff being sent to Sudbury, (although he soon returned) and F/O Sellers posted to Fort William. With quarters unavailable, all new officers and airmen boarded in private homes. Squadron personnel continued to live at home, reporting for duty each day at 0800 hours.

Since many airmen trained in trades other than their official ones, everyone had to be trade tested again. Because of the testing, the Squadron found it had too many carpenters, and not enough coppersmiths (the air force term for metal workers). The men spent many days in lectures, relearning their

THE 1939 ROYAL VISIT. KING GEORGE VI INSPECTS THE 112 SQUADRON GUARD OF HONOUR AT THE CPR STATION IN WINNIPEG

112 SQN OFFICERS - S/L LITTLE IS IN THE CENTRE OF THE PHOTO

trades. This turn of events irked Art Johnson, a fitter who had joined the squadron in 1936. "We didn't know from nothing as far as the (Permanent Force) was concerned. We were absolutely idiots. So they had us in lectures, they stripped us of our grouping. We had to re-trade test to get our groupings back."[14]

Major-General Andrew McNaughton, the newly appointed commander of the 1st Canadian Division, arrived from Ottawa on 8 November 1939. Before heading overseas, he went to visit his hometown of Moosomin, Saskatchewan courtesy of 112 Squadron. S/L Little[15] flew McNaughton in Avro 184, while F/O Moncrieff and Corporal Klaponski carried his luggage in Avro 266. Taking off after supper, they stopped in Brandon before carrying on to Moosomin. Since Moosomin did not have an airport, and because it was already dark, Frank Klaponski relates, "there were cars circled in this farmer's field, they had their headlights on to indicate where to land.

So the two aircraft landed in this field, and the General's people, I guess his closest family, picked him up and took him away." Moncrieff and Klaponski drained the oil from the aircraft, and kept it warm at a nearby farmhouse. The next day, the crews returned McNaughton to Winnipeg.

Now with the country at war, 112 Squadron left Winnipeg for the first time in the unit's history. On 30 January 1940,

112 SQUADRON CORONATION PARADE ON 12 MAY 1937

the Chief of Air Staff, A/V/M Croil, ordered the Squadron to Ottawa. They were to arrive on 7 February, with all Squadron personnel, including attached RCAMC, making the move. The four Avros, 184, 224, 266 and 267 were shipped by rail to the Technical Training School in St. Thomas Ontario, while Tiger Moth 248, plus all weapons, remained in Winnipeg at 2 Equipment Depot. The Puff Target Trainer was stored at the Equipment Depot, "pending return of squadron from active service."[16] All other equipment, tools and vehicles were to accompany the unit to Ottawa.

Rumours flew on 1 February. The movement order arrived the next day and packing up was "well under way thirty minutes after order received" according to the squadron diary. "Details worked all night resulting in practically all packing finished this morning."[17] The unit's mess property was consolidated in a private storage company in Winnipeg. With the purchase of tickets on Canadian Pacific Railway (CPR), and the filling of boxcars, the Squadron prepared for the move to the Nation's Capitol.

As the Squadron prepared to move on 5 February, some last minute confusion caused excitement. Two aircraftmen were released as medically unfit, but because their records were on the train, the paperwork had to wait until the Squadron arrived in Ottawa.

All personnel who were below medical category "A" were to stay in Winnipeg, and unfortunately, this included the chief clerk. Nevertheless, he went with the Squadron to Ottawa, because "other clerks are not capable of taking over."

At 1800 hours, as the Squadron formed up at Winnipeg's City Hall, a large crowd of wives, relatives and friends gathered and the roll was called. The mayor came forward to present an air force ensign to the squadron, while each member of the unit received a City of Winnipeg pin. His Worship then stated that he was happy that wherever it went, 112 Squadron would be known as the "City of Winnipeg" Squadron. The Squadron marched down Main Street to the train station. "All troop movements were supposed to be completely secret, but amusingly, when we marched through Winnipeg to the station, it was more like a home town send-off, with relatives and other people lining the curb, and running along with us."[18]

112 SQUADRON DEPARTS FOR OTTAWA VIA TRAIN – LEFT TO RIGHT LOCKE, S. SLAUGHTER, SCOTT, STEVENS, JOBIN

THE WESTLAND LYSANDER AT ROCKCLIFFE

After arriving in Ottawa, the Squadron took over the Lysanders left behind by 110 Squadron when they went overseas.

The Westland Lysander, one of the first short take-off and landing (STOL) aeroplanes, had the capability of operating from a rough field. Originally built for army co-operation, it became well-known for its support of resistance forces on the Continent.

Training began almost immediately, under the supervision of the School of Army Co-operation at RCAF Station Rockcliffe. Within only two weeks, instructors gave the Squadron Saturday afternoons and Sundays off, due to the high standard already achieved.

There were occasional breaks in the normal routine. The Governor-General, Lord Tweedsmuir, died suddenly on 11 February, and the Squadron provided a guard for the funeral party on February 14th.

110 SQUADRON LYSANDER BEING SERVICED, 18 JANUARY 1940, ROCKCLIFFE

LYSANDERS AT ROCKCLIFFE ON 4 APRIL 1940

March 9th proved an eventful day as 112 Squadron suffered its first serious crash. P/O A.E. Cannon got lost while doing circuits in poor weather, and came down at Alexandria Bay, New York, over 70 miles away. The aircraft was written off, with the pilot spending a few days at the hospital in Kingston. Fortunately no injuries occurred to the observer,

THREE SQUADRON AIRMEN - LEFT TO RIGHT, DAVY TOTTLE, JIM DUGUID, OWEN GRIFFITHS ON 27 FEBRUARY 1938. TOTTLE AND DUGUID ARE WEARING THE HIGH COLLAR UNIFORMS OF THE TIME.

Leading Aircraftman (LAC) Duguid. The same day, a trainee air gunner, Aircraftman Second Class (AC2) Morrison, put several rounds through the tail of the target tow plane.

As spring approached, flying conditions at Rockcliffe deteriorated. Training was cancelled several times with either the runway or the weather unsuitable. Lysander 429 flipped on landing on 14 March, when the pilot, P/O R.E. Chandler, hit deep snow. The Lysanders were moved to Uplands for a month so that training could continue.

112 completed its training at the school on 5 April. Flying, drill, and ground training continued as before, now under the control of the Squadron. Continuing poor weather, along with low aircraft serviceability rates cut into the flying hours.

The Squadron anticipated an early transfer to Britain, but events accelerated with the German invasion of France and the Low Countries on 10 May. Four days later, some personnel were warned for overseas service, however the situation changed so rapidly that by 18 May the entire Squadron started packing. The advance party, comprised of fourteen officers and eighty-five other ranks, embarked on the *SS Duchess of York* from Montreal two days later. This group included reinforcements for 110 Squadron. As the main party continued its preparations at Rockcliffe, the station's Lysanders were ferried to Toronto by Squadron pilots on 23 May. A week later, the squadron diarist noted the chaos:

112 SQUADRON LAC LEAVING FOR DEPLOYMENT OVERSEAS

THE ADVANCE PARTY DEPARTING ROCKCLIFFE FOR THE UK IN 1940.

Pay parade at 1000 hours was the cause of more commotion. Assignments of pay, dependent allowances - all served to provide the new paymaster who is accompanying us overseas (Lieut. Cann) with a grand headache. Sudden wealth caused a few of same that night. A muster parade was held at 1400 hrs. with full kit and gas masks. The men looked quite business like with their packs et al. There is a general feeling of anxiety to move to the docks. When will the order come?

The main party took a train for Halifax on 7 June, where they boarded the *SS Duchess of Atholl*. After four days, the ship finally sailed on 11 June, reaching Britain on 20 June.

112 Squadron played a small role in the RCAF in the 1930's. Being under-funded and with no modern equipment, the

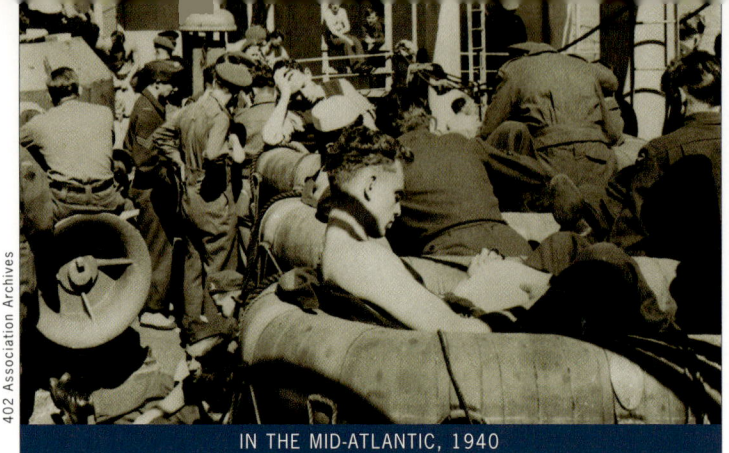

IN THE MID-ATLANTIC, 1940

Squadron operated under severe constraints. Nevertheless, officers from 112 held prominent positions in the wartime and postwar RCAF and the Squadron's training proved useful as the Air Force expanded. Other ranks easily stepped into full-time positions. With the start of the war, the identity of most auxiliary squadrons quickly disappeared, and became the lineage of new and vastly different fighting units. 112 Squadron, however, clung to its roots as the "City of Winnipeg" Squadron.

NOTES

1   W/C F.H. Hitchens, *Air Board, CAF and RCAF 1919-1939*, unpublished manuscript, page 221, quoting Hansard.

2   Another aim was the publication of the journal *Canadian Aviation*.

3   "Vancouver's Weekend Warriors", *High Flight* Volume 1, Number 1, page 25.

4   Privy Council Minutes PC2198. Both McNaughton's memo to the MND and the MND's proposal to the Governor-In-Council use the same phrase.

5   Gordon to McNaughton, 14 December 1932, National Archives of Canada (NAC), RG24, Vol 4951, File 895-9/112.

6   Later A/V/M Guthrie.

7   Air Force Orders, NAC RG24 Vol 3502 File 898-1-106.

8   Memo, Senior Air Officer to Deputy Minister, 09 May 1938 NAC RG24 Vol 3502 File 898-1-106.

9   Ibid.

10   Letter CO 111 Sqn to HQ MD 11 13 July 1938, NAC RG24 Vol 3502 File 898-1-106.

11   112 Squadron Report on Annual Summer Training, NAC RG 24 Vol. 3384 File 450-9/112.

12   W.A. Curtis was a pre-war auxiliary member commanding 110 Squadron from 1935 to 1938.

13   NAC RG24 Vol 3502 File 898-1-106.

14   Interview with Art Johnson.

15   S/L Little took command from S/L Crabb on 1 October 1939.

16   Movement Order 30 January 1940, NAC RG24 Vol 4951 File 895-9/112.

17   3 February.

18   Ken Smith "Manning Pool: Training Airmen in the First Year of World War Two" CAHS Journal Vol 27, No 1.

402 SQUADRON HURRICANES CUE UP FOR TAKE-OFF

John Griffin Collection

The Second World War

Stephen M. Fochuk

June to July 1940

When the *S.S. Duchess of York* pulled into the harbour at Liverpool on 20 June, little did the men know that their intended role of supporting the Canadian Army in France was about to change. The British Expeditionary Force (B.E.F.) was in full retreat and the bulk of the Canadian Army could not get to the continent to support it. The evacuation of the last of the B.E.F. from Dunkirk took place on 3-4 June, and although some units lingered on for another week in sectors along the French coast, the Battle of France, as it became known, ended during the first week of June.

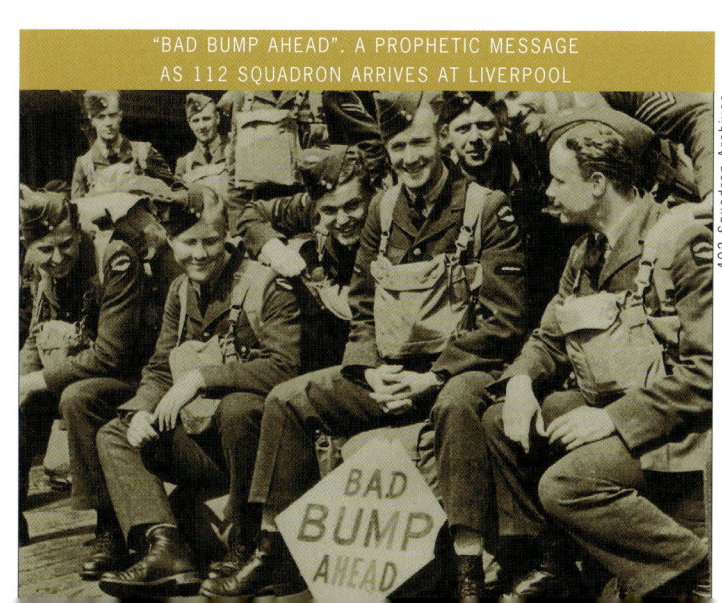

"BAD BUMP AHEAD". A PROPHETIC MESSAGE AS 112 SQUADRON ARRIVES AT LIVERPOOL

402 Squadron Archives

CHAPTER 2

MAP OF 402 STATIONS IN THE UK

they met up with the advance party and boarded a bus for their final destination, Royal Air Force (RAF) Station High Post, Wilts. That very day, the French Government capitulated after signing an armistice with Germany.[2] Now, only the English Channel kept the German Blitzkrieg from rolling over Britain.

The Squadron busied itself over the next few weeks by taking care of enormous amounts of paperwork, and settling into their new surroundings. Personnel were directed to beef up the ground defences around the station.[3] Fear of a German invasion created a state of high alert in Britain and rumours were rife that paratroopers might precede an invasion. In the midst of this, the first three Westland Lysander aircraft were delivered on 23 June, with the rest trickling in over the next few weeks.[4]

In essence, No.112 and her sister squadron, No.110, had trained for a role which was no longer required.[1]

After disembarking from the *York*, the Canadians made their way to a camp at Weston where they spent the night. The next day (22 June) they proceeded to Salisbury, Wiltshire, where

112 SQN LYSANDERS ON A TRAINING FLIGHT

WE STAND ON GUARD

ONE OF THE INTERIM HAWKER HIND AIRCRAFT THE SQUADRON USED PRIOR TO THE AVAILABILITY OF THE LYSANDERS

The arrival of the Lysanders seemed to be a sign of progress, but the discovery that both parachutes and airframe/engine logbooks were missing caused some exasperation within the Squadron. There was no choice but to ground the Lysanders until the delivery of the appropriate equipment and documentation. In the meantime, several Hawker Hinds, (a pre-war two-seat biplane) were procured and used as interim equipment.

The Squadron had their first taste of living in a war zone when, in the early hours of 25 June 1940, the air raid sirens wailed for several hours in anticipation of a raid that never materialized. The missing logbooks and parachutes finally arrived on 6 July, and flying began.[5]

The Germans wasted little time launching their attack on Britain. The Channel Islands fell on 30 June and the Luftwaffe air assault, named "The Battle of Britain" by Winston Churchill, began on 10 July.[6] Now, literally on her own, Great Britain suffered a steady pounding from the sky over the following months. In order to defend herself from the onslaught, the Country urgently needed fighter pilots and fighter squadrons.

Had Germany launched a sea-borne assault, 112 Squadron would have been used to attack enemy along the southeast coast of England. A letter from the Air Officer Commanding (AOC) of the RCAF in England, Group Captain (G/C) F.V. Heakes to the officer commanding No.112 Squadron, sheds some light on this matter:

112 SQUADRON HALTON, BUCKS - LORD ROTHCHILD'S ESTATE – BACK ROW L TO R: SMOKIE MACLAUCHLIN, RAY BOURGEAN, ALEX SCOTT, MARCEL JOBIN, BUD GAYFER – CENTRE ROW: MUSH HIGGINS, BOB FALLIS, JOHN MAVILLE, SPUD MURPHY, ERLE S. MILLER – FRONT ROW: ? CHILDS, BERNIE VILLENEAUVE, LUDGER NAVLLES?, HARVEY FACELLE?, STAN REID

It may be necessary, should an emergency arise, to employ part of your squadron on very short notice on active operations. You will, therefore, place in one flight sufficient of your best-trained personnel, including pilots and maintenance personnel, to operate six aircraft as a special Service detachment.

The emergency referred to in para 1, has particular reference to attempted invasion of this country. The employment of such a detachment, however, will only be undertaken in an extreme emergency.[7]

Fortunately that emergency never arose. Had they been called upon, 112 with their Lysanders probably would not have faired well against the battle-hardened Luftwaffe pilots and their state-of-the-art Me109s. However, the Germans postponed their invasion, dubbed Operation Sea Lion, having failed to achieve the needed air superiority over Britain. According to official British sources, the Battle of Britain ended on 31 October.[8]

With the coming of winter and shorter days, the Luftwaffe's daylight attacks began to subside, giving the RAF a much-needed respite. During this period, Fighter Command expanded its single-engine fighter force, enabling the RAF to prepare for the spring or summer and the expected renewal of the Luftwaffe's assault.

112 Squadron became part of this expansion under S/L W.F. Hanna, CBE, who became O.C. (Officer Commanding) on 16 August 1940. William Hanna, who had seen action in the First World War as an observer with 48 Squadron Royal Flying Corps, and as a gunner, had one victory to his credit.

The first step for the Squadron, as part of Fighter Command's expansion, involved relocation to a quiet sector of Britain where the conversion to fighters could take place without much chance of enemy interference. Along with the change of location, the unit became No.2 Canadian Squadron on 11 December.

A 112 SQUADRON LYSANDER MK I

S/L W.F. HANNA, CBE, STANDING IN THE CENTRE

IN ONE OF THE MOST FAMOUS PHOTOS OF THE SQUADRON, PILOTS SCRAMBLE FOR THEIR HURRICANES, WHILE THE SMILING GROUNDCREW WAIT WITH CHUTES AT THE READY

In mid-December, they were sent to No.12 Group's RAF Station Digby for conversion to a modern day-fighter, the Hawker Hurricane Mk I.[9] The Hurricane will always be remembered as the backbone of Fighter Command during the Battle of Britain. Indeed, in the summer of 1940, the RAF possessed almost twice as many Hurricanes as it did Spitfires. The fighter was a rugged, easily repaired aircraft, which provided her pilots with an excellent gun platform. The first of

HAWKER HURRICANE COLOUR PROFILE

the Squadron's Hurricanes arrived on 14 December and training soon began. Two days later, the unit officially became classified as a Single Engine Fighter (S.E.F.) squadron.[10] The first solos took place on 18 December and four days later, all pilots had flying time in the new fighter.

Although the war raged on, the Squadron stopped to observe Christmas. In keeping with an Air Force tradition that continues to this day, the officers held an open house at their mess for the sergeants. The officers then proceeded to the airmen's mess, where they served the men an excellent dinner. Later, gifts of cigarettes, candy, razor blades and other items of comfort were distributed.[11]

No.2 Squadron's diarist finished the year with this entry: "Normal routing this last day of 1940. Here's to victory in 1941."[12]

CHRISTMAS 1941 AT WARMWELL WITH THE SQUADRON'S GROUNDCREW AND AIRWOMEN

January to February 1941

S/L GORDON MACGREGOR

1941 began with new O.C., S/L G.R. McGregor, DFC, taking command of the Squadron. MacGregor, a 39-year-old veteran of the Battle of Britain, was well qualified to lead the new fighter squadron. With the urgent need to make his squadron operational in the shortest possible time, McGregor had the men carry out local reconnaissance flights, perform battle climbs to 30,000 feet and practice formation flying. Snowy weather caused unwanted delays in the training program. For instance, no flying took place on the 19th due to an all-day snowstorm which kept everybody, including the pilots, busy removing the accumulation from the runways and around the station.

When not flying (or shovelling), the pilots occupied their time with the Link trainer, attending lectures on fighter tactics, radio/telephony (R/T) procedure, reflector gun-sight, and armament studies.

February did not bring an improvement in the weather, but the Squadron did manage to get some time in the air. Not all of the practice flights were dull and mundane; for example on 25 January 1941, while engaged in a battle climb to 25,000 feet, P/O G.A. Russell lost consciousness. Upon recovery, he found the hood, emergency panel and part of the fuselage of his Hurricane missing. Russell crash-landed on the aerodrome at Digby, walking away uninjured. Further examination of the Hurricane pointed to a disintegrated airscrew (propeller) which probably caused the missing pieces and torn fuselage.[13]

WARRANT OFFICER GRAHAM "CARP" CARPENTER (FRONT RIGHT) AND HIS INDEFATIGABLE TECHNICIANS

THE SERGEANTS' MESS AT SOUTHEND-ON-SEA ESSEX, 41 ANN BOLEYN STATE – L TO R: FRANK KLAPONSKI, JIM DUGUID, G.D. ROBERTSON, GRAHAM CARPENTER, G. MURCHIE, D. ANDERSON

At the end of February, pilots were being "declared day operationally trained."[14] On 28 February 1941, A/V/M R.E. Saul, AOC No.12 Group, visited the station and spoke to the pilots. After Saul's speech, 2 Squadron's officer commanding, S/L G.R. McGregor, informed the A/V/M that all but two pilots were "declared day operationally trained"[15] and the Squadron could officially be classified 'day operational'. McGregor received a congratulatory letter from Saul two weeks later.

> My Dear McGregor,
>
> I have noted with considerable satisfaction the rapid progress made by No.402 (Canadian) Squadron, since the date of its formation in this Group.
>
> The keenness and fine spirit existing throughout the Squadron is most noticeable, and it is largely responsible for the Squadron becoming operationally trained in such a short period.
>
> one of our formost [sic] when the chance of getting into action against the enemy comes their way.
>
> I wish to congratulate you on your fine leadership and on the manner in which you have built up this fine Squadron, and you have every reason to be proud of your Command.
>
> Please convey my appreciation to all ranks, and I wish them the very best of luck in their future activities.
>
> Yours sincerely, R.E. Saul

March to October 1941

On 1 March, the newly declared 'operational' squadron received yet another designation, and was henceforth

THE SQUADRON AT DIGBY

known as 402 Squadron. In order to avoid confusion with the RAF Squadrons, the RCAF had been allocated the block of squadron numbers from 400 to 450.[16] 402 consisted of two flights, 'A' and 'B'. The first nominal roll for the squadron read as follows:

'A' Flight - F/L V.B. Corbett, F/Os G.G. Hyde, C.T. Cantrill and J. Saint-Pierre. P/Os J.A. Thompson, F.W. Kelly, D.J. Smith, F.B. Foster, R.R. Walker and W.H. Pentland. Sgts. G.D. Robertson, G.W. Walker and H.L. Palton.

'B' Flight - F/L E.M. Reyno, F/Os T.B. Little, R.E. Morrow and H.J. Findlay. P/Os L.V. Chadburn, H.S. Crease, R.R. Gillespie, N.H. Bretz, J.E. Walker and D.L. Ramsay. Sgts. N.B. Trask, D.W. Jenkin and K.B. Handley.

Immediately, the newly renumbered 402 Squadron achieved operational status, with 'A' Flight adopting a state of readiness. The only operational flight that day proved uneventful. The next day, 'B' Flight had their turn and for the rest of the month the flights alternated from day to day.[17]

With the winter quickly fading away and the threat of a renewed German offensive looming, Britain remained on a state of alert. Meanwhile, Fighter Command brought the Canadians into the fray slowly, with 402 spending the better part of March on readiness and defensive duties. Usually, these were patrols but they did occasional sweeps[18] to France as well.

402 SQUADRON, JULY 1942 - INFO INCOMPLETE — WO HAMILTON, SGT L.M. CAMERON, SGT A.M. SKINNER, FSGT O.R. BROWN, SGT G.C. MCGRAW, SGT E.J. ROSS, F/L N.B. TRASK, P/O J.R. SCOTT, MAGEE, S/L R.E. MORROW, P/O G.N. KIETH, FSGT N.A. KEENE, F/L N.H. BRETZ, P/O I.G. KELTIE, P/O H. RUSSELL, FSGT K.W. BIRD, P/O J.C.U. BAYLY, SGT G.D. CAMERON, F/O YEANDLE (ADJ)

March 23, 1941 marked a significant date for the Squadron. The previous day, 402 had been sent to RAF Station Coltishall, located near the Norfolk coast, to reinforce a sweep. With persistent fog and rain preventing them from both carrying out their assigned task, and returning to Digby, they spent the night at Coltishall. On the 23rd, the Winnipeg Bears[19] were scrambled no less than seven times, with one sortie yielding some action. F/O R.E. Morrow and Sgt K.B. Handley took part and in their official account, as told to Squadron Intelligence Officer, P/O Brown, explained what transpired:

SGTS "BUTCH" HANDLEY AND GRAHAM ROBERTSON GOING OVER THE LATEST OP

Black Section of No.402 (Canadian) Squadron, two Hurricanes flown by F/O R.E. Morrow and Sgt Handley, Black One and Black Two respectively, were ordered to take off from Coltishall at 1243 hours and fly on a course of 070 degrees at 15,000 feet. At 3,000 feet the course was changed to 020 degrees, and at this point, in cloud, Black Two got separated from Black One, and was unable to regain contact or take any part in the ensuing combat. The cloud density was 8/10 extending from 3,500 to 8,000 feet (10/10 being completely overcast). As the sky was clear above 8,000 feet, the Controller instructed Black Two to circle until rejoined by Black One. While on his way to rejoin Black Two, Black One was directed on to an enemy aircraft in the vicinity.

While flying at 16,000 feet and at 250 mph Black One sighted the enemy aircraft 1,000 feet below him, ahead slight to starboard traveling north at approximately 200 mph. This was at a point approximately 35 miles North East of Cromer. Black One turned about 20 degrees to starboard and made a stern quarter attack on port side of the enemy which he identified as a Ju.88, firing three second burst at 400 yards closing to 200 yards.

The enemy replied with a short burst from his upper rear guns, and then went into a steep spiral dive. Black One followed him down to 7,000 feet firing four more bursts at ranges of from 300 to 150 yards. The enemy replied with one very short burst and then ceased firing, and no further return fire was experienced. The enemy entered 6/10 the cloud at 7,000 feet and Black One lost contact but went down through the cloud at about 2,000 feet and observed 3 or 4 disturbances in the water, which made him think that the Ju.88 had jettisoned its bombs. Black One made a further search both above and below cloud but was unable to regain contact. Both Black One and Black Two landed at Coltishall at 1335 hours. During the combat F/O Morrow noticed what appeared to be smoke issuing from the enemy's port engine.

This marked the first time the Squadron had made contact with the enemy. Celebrations ensued when the detachment returned to Digby later that day.

"WE STAND ON GUARD" – A 402 AIRMAN KITTED OUT FOR SENTRY DUTY

Fighter Command spent the next several months on a defensive footing trying to guess what the enemy's plans were. It was thought another assault could happen at any time, but probably not until the spring or summer brought improved weather.

September was thought to be the probable cut-off date for a German aerial campaign; meanwhile, the Luftwaffe remained active. The weather of course affected them too, but when opportunities were available, the Germans bombed both civilian and military targets.[20]

NORTH LINCOLN HEATH HOUSE –
LEFT TO RIGHT: F/O HARRY CREASE, F/O BOB MORROW, SGT GRAHAM ROBERTSON, F/O "ROCKY" ST. PIERRE, SGT JERRY WALKER, F/O JIM THOMPSON, F/L VAUGHN CORBETT

On 14 April, S/L G.R. McGregor was promoted to Wing Commander (W/C), and in taking command of the Digby Wing, became the RCAF's first Wing Leader.

S/L V.B. Corbett, became the Squadron's new O.C., however tragedy marred the change of command day. A Ju88 burst from the low hanging clouds over Digby, dropping four bombs on the aerodrome. Two of them narrowly missed the wireless towers and station post office, while one landed in a nearby field and the other on a road. Sadly, the Squadron lost its first member as a result of enemy action. Two airmen, Aircraftman First Class (AC1) J.F. Tighe and LAC J.E. Owen were driving to Ashby Hall, located approximately two miles from Digby, when a bomb landed beside them. According to witnesses, the resulting blast blew their vehicle thirty feet into the air. It landed on its nose twenty-five feet from the road. Both men were thrown from the vehicle and Owen died from massive head injuries. Tighe was found conscious, and although bleeding profusely with broken vertebrae and a pierced lung, survived. James Ernest Owen, a 20-year-old from Peterboro, Ontario is buried in the Scopwick Church Burial Ground in Scopwick, Lincolnshire.[21]

S/L Vaughn Bowerman Corbett had been one of the first Canadians injured during the war. Shot down in flames while with No.1 Squadron RCAF, he came to command 402 after a lengthy stay in hospital recovering from his burns. He survived the war only to die in a 1945 flying accident while Station Commander at Bagotville, Quebec.[22]

402 HAWKER HURRICANE MK IIB, BE485, CODED AE W, ARMED WITH A PAIR OF 250 LB GENERAL PURPOSE BOMBS

The day after the change of command, 15 April, 402 accompanied two Spitfire Squadrons, 234 and 65, on a Rodeo[23] along the French coast. Led by the newly promoted W/C G.R. McGregor, the Rodeo proved uneventful but not insignificant. Indeed, it established 402 as the first RCAF unit to carry out an offensive operation over enemy territory.[24]

Towards the end of the month, the Squadron was re-equipped with the latest model of the Hawker Hurricane, the Mk IIB. The new Hurricanes were phased in over the next several weeks. During the month of May, 402 underwent night flying training during the hours before sunrise and after sunset. This type of training coincided with improving weather, and the renewed prospect of a German assault on Britain. The possibility of an attack during the cover of darkness had become a distinct possibility.

Due to its nature, night flying had its share of incidents. During the early hours of 4 May, F/O G.G. Hyde collided with another aircraft while taxiing. After being given the go-ahead to begin his rollout onto the runway, he ran into a stationary aircraft parked just off the perimeter strip. Both aircraft were damaged, Hyde's being classified as category (Cat) AC.[25]

On 8 May, six pilots stood by on a state of readiness for night operations. They took off at 0000 hours to patrol the Derby and Sheffield areas. Two enemy aircraft were sighted at some distance but they disappeared into the darkness. The flight returned to base at 0200 hours where a lighted flare path aided their landing. S/L Corbett was on final approach when he was attacked by what was thought to be a Junkers Ju88. Corbett

ARMOURERS HANDLING A BOMB AT WARMWELL – THE CAPTION FOR THIS IMAGE AS IT APPEARED IN CANADIAN NEWSPAPERS DURING THE WAR, READS: " 'BUNDLE OF GRIEF' BEING PREPARED FOR THE NAZIS. THESE TWO ARMOURERS HAVE FUSED THIS BOMB AND ARE SECURING THE TAIL FIN IN ITS PROPER POSITION. THEN THEY WILL LOAD THE SENSITIVE 'EGG' UNDER THE WINGS OF THE VICIOUS 'HURRI-BOMBER' BEHIND."

immediately climbed away, leaving the enemy aircraft to pass beneath him without scoring a hit. Still, the officer in charge of the flare path received minor injuries from the bullets, and an airman broke his leg tripping in the rush to take cover.

In the early hours of 22 June 1941, Germany launched Operation Barbarossa, the invasion of Russia, with consequences affecting the Squadron. First, it gave the RAF more time to build up and expand its fighter force, and secondly, the pace of the offensive carried out by Fighter Command increased in an effort to relieve some of the pressure from the Eastern front. "Apart from political aims, the object of the fighter offensive of 1941 was two-fold. Enemy ground targets, ships and above all aircraft were to be destroyed and was also hoped that by their destruction, the enemy would be prevented from withdrawing units from the west and sending them to Russia."[26] Lastly, it extended the resources of the Luftwaffe which now fought on three fronts: Western, Eastern Europe and Mediterranean. With Operation Barbarossa in the east, Fighter Command did not expect the Luftwaffe to launch another offensive against them unless either of two things happened. One encompassed German territorial gains during the push toward Moscow. The other was that the Russians would simply surrender. The British surmised that these events might happen in the spring or quite possibly in the summer of 1942.

MARCH 1941 AND THE GROUNDCREW AT DIGBY GET CARRIED AWAY WITH "KILL" MARKINGS. L TO R: J. GILLESPIE, CPL T. RYLAND, A.J. CAMERON, J. JOBIN

DIGBY 1941 - HURRICANE IIB BE489, AE Q WITH NOSE ART "BUZZ THE FALCON" ON FSGT "BUTCH" HANDLEY'S AIRCRAFT – CPL T. RYLAND LEFT, AND CPL L.R. GRAHAM

SGT DUGUID LEAVING FOR FAIRWOOD COMMON

The day after the opening of Operation Barbarossa, 402 moved to RAF Station Martlesham Heath. Here they spent a few weeks carrying out defensive patrols before being relocated to Ayr, Scotland on 10 July. The move allowed 402 to train in a quiet sector and prepare for the start of Fighter Command's new offensive. Operationally speaking, the Squadron maintained a state of readiness while at Ayr. When not on alert, they continued to train.

Ready for action once more, 402 moved to RAF Station Southend/Rochford, located along the east coast of England, on 19 August 1941. The move brought them closer to the action as part of No.11 Group.

Members were called upon to take part in a sweep on 23 August, however inclement weather cancelled the show before they got off the ground.

The next day, S/L V.B. Corbett and F/O F.B. Foster flew to RAF Station Duxford, Cambridgeshire, where they carried out test flights with a modified Hurricane. This aircraft was the recently developed 'Hurribomber', normally equipped to carry two 250-lb bombs, one under each wing.[27] Since the modified Hurricanes were destined for 402 Squadron, both Corbett and Foster became very familiar with the aircraft.

MARTLSHAM HEATH 1941. AL SCOTT (LEFT) AND FERN DELANGESE

On 27 August, the Squadron took part in a Circus[28] to the Lille, St-Omer area, with 13 Bristol Blenheims (a twin-engine light bomber) and Spitfires from the North Weald Wing. S/L Corbett's Hurricanes had a rendezvous with the Spitfires at 0712 before meeting the bombers thirteen minutes later over Manston, Kent.

S/L Corbett's aircraft suffered a hit from the intense flak encountered while crossing the French coast, but because his Hurricane seemed responsive, he chose to remain with the Squadron. Although the target had not been sighted, F/L Morrow distinctly heard a "bombs away" call through his R/T. During the return trip, an Me109 dived just ahead of the Squadron firing tracers and shooting down Sgt D.W. Jenkin. Fortunately Jenkin survived and became a Prisoner of War (PoW). Then, mayhem broke out about fourteen miles northwest of Calais over the Channel when F/O H.S. Crease witnessed a Spitfire collide with F/L T.B. Little's Hurricane.

The devastating impact cut the Hurricane's tail off about three feet aft of the cockpit and although Little managed to bail out, he did not survive. F/L Little likely collided with F/L J.C. Martin of 222 Squadron, who also lost his life. A Battle of Britain veteran, Thomas Burgess Little, 24 years of age from Montreal, Quebec, has no known grave. His name is inscribed on the Runnymede War Memorial, Englefield Green, Egham, Surrey, England.[29]

Fighter Command's offensive was in full force by the time the Squadron relocated to Southend, where they were soon taking an active role in supporting Circus operations. These Circuses continued to involve 402 well into the first weeks of October.

No sooner had the Winnipeg Bears become comfortable in their new surroundings; they were removed from the active area. Fighter Command had decided to restrict operations in order to conserve their aircraft for the 1942 season (this plan slowly came to fruition with the improved weather and increased daylight). Coinciding with the slowing down of the offensive, a new task/responsibility was placed on the shoulders of Fighter Command. The Channel Stop[30] a responsibility of No.2 Group for the first half of the year, had been handed over to Fighter Command on 8 October due to the heavy losses the Blenheim bombers were suffering. The Channel Stop was to be carried out with support from the new Hurribomber and thus included 402.

November 1941 to March 1942

The end of September saw the arrival of the first new Hurribombers, and practice bomb-drops took place on 2 October. Apart from being placed on readiness, 402 spent the rest of the month carrying out training flights with the new aircraft and getting used to the added weight of the two bombs.

The Squadron's first operational mission with the Hurribomber took place on 1 November 1941 as a part of Ramrod 3A.[31] Eight Hurricanes with a fighter escort attacked the Berck-sur-Mer aerodrome. The mission went according to plan and the after action report explains what transpired:

HURRIBOMBER DELIVERS A 250-POUNDER

Rendezvous was made at Rye under 500 ft at 1220 hrs., escort to be; Close 303, Top 315, Rear 308 Squadrons. One escort squadron was seen to go ahead off Rye, the two remaining squadrons maintained contact and orbited 5 miles off the French Coast, and provided escort home. The 8 Hurricanes approached the Coast South of Berck-sur-Mer in line abreast at 0 feet. F/O F.W. Kelly bombed 35 railway wagons including tankers, on a siding to a factory near Rang de Fliers from a height of 30 feet. He gave a 2 second burst at a Bofor [sic] type gun nearby.

The Squadron then proceeded at very low altitude turning towards Le Canche river just South of Montreuil. Here, F/L R.E. Morrow dropped his bombs from a height of 10 ft., on a railway junction, a miniature railway running from the main line, shooting down a German solder [sic] who was aiming in his direction with a small arm weapon shortly after.

P/O N.B. Trask dropped his bombs in the centre of a 3 gun post, fired a 3 second burst into a signal box near Bentin and saw chunks falling off the box, finally shooting up a gun post on the sand dunes as he floew [sic] out North of the river. The Commanding Officer, Squadron Leader V.B. Corbett, flying down the river from Montreuil, bombed the railway river bridge to Etaples from 20-30 feet, the spout in the centre of the river bridge to sway and buckle toward the West. He later shot up with a 4 second burst, a Nissen hut on the S.W. corner of Paris Plage drome. F/O F.B. Foster bombed the same bridge as his C.O. from 10 feet, his bombs dropping on the further side. He later shot up a M.G. in N.W. corner of Paris Plage. The burst being seen by F/L Morrow. P/O N.H. Bretz bombed Railway between Villiers and Etaples from 40 feet. Result not seen. P/O L.S. Ford fired a 10 second burst along some 20 huts one mile North of Etaples and silenced a machine gunner, then bombed 12 barges drawn up on the mid on the East side of Le Touquet town, just within the ostuary [sic], from 10 feet. P/O D.J. Smith saw these bombs burst on the barges, he himself running in from over the town, dropped from 200 feet the bombs overshooting the barges.[32]

The Squadron returned to the same target three days later. During this attack, Sgt Handley noticed two snub-nosed aircraft in the southeast corner of the field. These

A CAPTURED FOCKE-WULF (Fw190)

were thought to be the Luftwaffe's potent new fighter, the Focke-Wulf 190 (Fw190).

Two days later, the Squadron relocated to RAF Station Warmwell, on the south coast of England, bringing them once again, closer to the action.

TEA TIME AT THE NAFFI WAGON, WARMWELL

On 26 November 1941, 402 carried out its first mission related to the Channel Stop. Earlier in the day, a ship of approximately 3000 tons had been reported just off Cherbourg Harbour. Four Hurricanes in two sections took off at 1315 hours. Escorted by Spitfire Squadrons Nos. 234 and 501, the formation set a course for C. De la Hague. Reaching their destination, they turned west and flew along the coast to Cherbourg. The area seemed vacant of any shipping, but when three enemy aircraft were spotted in a wide formation with another four behind them, the squadrons decided to return to base and landed at 1420 hours.[33]

The next two months were quiet ones for 402, with the lion's share of the time being spent on training flights. The only operational flying involved providing cover to convoys plying the Channel. Poor weather, shorter days, and the need to conserve the fighter force all contributed to the lack of operational flying.

After receiving a DFC for his outstanding leadership of the Squadron, and a posting to RCAF Overseas Headquarters, S/L Corbett handed command to S/L R.E.E. Morrow, another commanding officer promoted from within the Squadron's ranks.

Sightings of enemy aircraft were rare during this period, however on 2 January 1942, the Squadron ran into a pair of Me109s:

> Two Hurricanes Mk II, Bombers 2 x 250 G.P. 11 second delay each of Red Section, (P/O Ford and Sgt. O'Neill) 402 Squadron R.C.A.F. left Warmwell at 1045 hours on a shipping reconnaissance patrol. Red Section steered a course of 160 degrees at 0 feet ten or fifteen minutes, then pulled up to seek

S/L BOB MORROW

> cloud cover which was 10/10 at 5,000 ft. On reaching a height of 3,000 feet, a further bank of 10/10 cloud was seen at 2,000 feet. Altering height, Red Section flew about 100 feet under this cloud bank until the French coast was sighted near Cap-de-Barfleur. On sighting the coast, Red Section changed course to 25 deg., and patrolled ten miles off the coast to Cherbourg, no shipping was seen in the Channel of Cherbourg Harbour. About eight miles off Cherbourg the Section was suddenly jumped by four Me.109E's, which came out of the

cloud. Red 2 who was flying below and behind his leader, called "109's, 109's" over the R/T. Red 1 was fired at, his port aileron being partly shot away by cannon fire, and damage was done to the port wing by machine gun fire. Red 1 immediately jettisoned his bombs and dived to port to elude the e/a's fire, pulling out, he noticed two e/a on Red 2's tail, getting the nearest of the two in his sights, he gave a one second burst at 100 yards range full deflection, but had to break off the attack without noticing any result, as the damaged aileron made the aircraft hard to manoeuvre. Red 1 was being fired at all this time and decided to seek cloud cover.

He was now at 1,000 feet and climbed and skidded several times with good success until he reached cloud cover at 2,000 feet, there being no further strikes on his aircraft.

After being in cloud a few seconds Red 1 dropped out again and was again attacked, he immediately pulled back into cloud and proceeded back to base. The last time Red 1 saw Red 2, he (Red 2) was in a shallow dive, banking left about 50 feet off the water with Glycol streaming behind.[34]

O'Neill's position was noted, and on their return to base the Squadron sought permission to undertake a rescue search. This was refused because an enemy R/T message indicated that the German air sea rescue services were taking all possible action. Unfortunately, FSgt O'Neill was killed when his aircraft crashed into the sea. Bernard Peter O'Neill of Timmins, Ontario, 23 years of age, rests in the Old Communal Cemetery, Cherbourg, Manche, France.[35]

On 11 February 1942, the Germans put Operation Cerberus into effect. This bold plan enabled three major German warships, the battle cruisers Scharnhorst and Gneisenau along with the heavy cruiser Prinz Eugen, to depart Brest harbour for their homeports in the north. The British had successfully been able to hold the ships in the French port up to this point. The operation, later known as the Channel Dash, occurred over the next three days. The Germans surmised that the British would never expect the ships to sail in broad

IN 2007, REAL AEROPLANE COMPANY IN THE UK CHOSE TO PAINT THEIR HURRICANES IN THE COLOURS OF MK IIB BD707 OF 402 SQUADRON. THIS AIRCRAFT WAS USUALLY FLOWN BY FSGT GRAHAM ROBERTSON. ON SEPTEMBER 18, 1941 HE SHOT DOWN ONE ME109 OVER CHANNEL, HENCE THE ONE SWASTIKA AS A KILL MARKING UNDER THE COCKPIT. HE WENT ON TO BE COMMISSIONED AND EVENTUALLY LED HIS OWN SQUADRON.

daylight during low tide. The British were completely unprepared, as witnessed by the quickly thrown together and ineffective attacks on the ships.[36] Seventy-one of three hundred and ninety-eight aircraft that tried to launch attacks against the German flotilla were shot down without a single bomb hitting a target. Deplorable weather and fierce anti-aircraft guns at Pas-de-Calais were key elements in the German's success of Operation Cerberus.[37]

Despite Scharnhorst hitting two mines and Gneisenau one, the ships made it to Wilhelmshaven much to the dismay and injured pride of the British.[38] Although 402 with their Hurribombers may have been able to make a difference, they were never called into the attack.

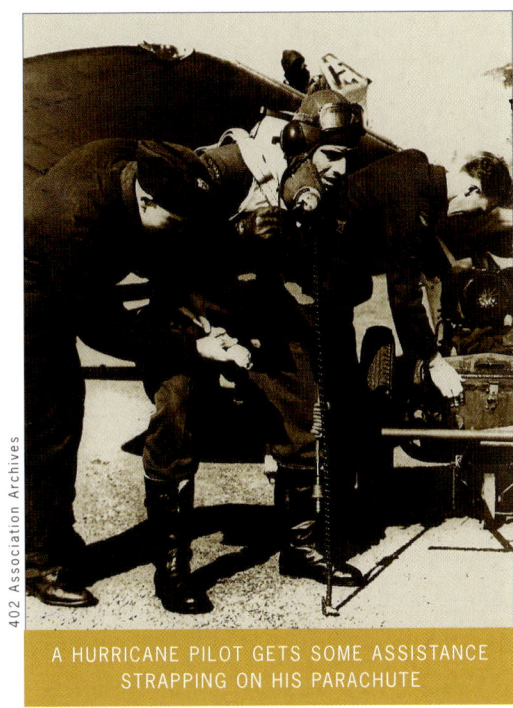

A HURRICANE PILOT GETS SOME ASSISTANCE STRAPPING ON HIS PARACHUTE

The Squadron was directed to Manston, Kent, on the 12th. Despite being a very active period, the only action the unit saw was initiated by the Luftwaffe. A pair of twin-engine Dornier bombers attacked Warmwell at 0845 hours, bombing and machine-gunning the aerodrome. Slight damage occurred and fortunately, only one person, a soldier of the Dorset Regiment, received minor wounds from splinters or fragments.

The Squadron stood by on readiness the morning of 16 February 1942. That afternoon, 402 received orders directing them to attack a small convoy of minesweepers off the Brittany coast. After refuelling and picking up a Spitfire escort at Perranporth, on the west coast of Cornwall, S/L Morrow led his six Hurricanes, each carrying two 250-lb bombs, on a low level attack. To the surprise of the Squadron, the minesweepers turned out to be heavily armed destroyers. At 50 feet above the waves, the Canadians bore down on their targets, paying little regard to the formidable fire they were facing. As a result of this action, two ships were badly damaged. One absorbed a hit from a single bomb and another was struck twice. Although badly damaged, both vessels managed to make the Port of Brest. Morrow's Hurricanes all returned to Perranporth; however one of them crashed on landing.

402 remained at Perranporth. The next morning at 0830, "A" Flight was at readiness, armed with bombs. A stand down was called, and at noon, the flight flew back to Warmwell. During the return trip, three aircraft of 'A' flight

engaged an equal number of enemy aircraft. The General Report of the incident details the action:

Three Hurricane Mk II bombers, two of whom were bombed up with 2 x 250 lb., General Purpose bombs of Yellow section, (Yellow 1 F/O Ford, Yellow 2, FSgt Emberg and FSgt Keene) left Perranport [sic] at 12:45 hrs for the purpose of returning to Warmwell. Shortly after take-off, FSgt Emberg noticed his port undercarriage light was not on, he reduced airspeed and it was some time before he was able to lock his undercarriage up by means of the hand pump. He had by this time lost sight of his leader in the poor visibility and not being very well acquainted with the coast in this area he decided to continue across the Channel from Dodmans Point to Start Point, flying at 1700 ft. R/T reception was very poor owing to a load buzzing interference. FSgt Keene who had fallen back with Yellow 2 formed up with him flying about three spans out and 50 yds to the rear.

After approximately 30 minutes Flying, Yellow 2 had not sighted Start Point and had also lost contact with FSgt Keene who stated that he had lost his leader in cloud approximately 25 minutes after formatting on him. After missing FSgt Keene, Yellow 2 began making gentle turns trying to locate him as Yellow 2 was doing this he noticed an Me.109F come out of cloud about 150 yards to his port on a converging course to his own. Yellow 2 dived about 50 feet and pulled up to deliver a quarter stern attack at 100 yards range and slight below the E/A using a 2 second burst at 2 rings deflection, at the moment of attack, the E/A was banked slightly to port. Yellow 2 saw hits along the fuselage of the E/A. The enemy then started a very gentle diving turn to port. Yellow 2 then turned and delivered a quarter stern attack on the E/A's port side giving a further 2 second burst at 150 yards. The E/A's angle of glide increased and started to turn to starboard, Yellow 2 again attacked this time with a 3 second burst from the quarter stern position on the starboard side at approximately 200 yards, at 2.5 rings deflection. The enemy a/c then went into a sideslip and crashed into the sea.

Yellow 2 came down to about 100 feet and watched the enemy sink, he then noticed that he was being attacked from dead astern by two E/A's which he later identified as Me.109F's.

Yellow 2 immediately started evasive action, weaving from side to side and then attempted to climb for cloud but seeing he would be a sitter by doing this, he turned and dived back towards the sea, later after evading a number of diving attacks by the E/A, he went into a steep climbing turn to approximately 1400 feet, then straightened out and climbed vertically into cloud cover which was 10/10 at 1700 feet.

Keeping in cloud cover, Yellow 2 set course of 350 degrees for the English Coast. Shortly after setting his course, Yellow 2 noticed his engine began to run rough and start to cut out. He then came out of cloud cover and continued on his course, Yellow 2 now started to call for an emergency homing and also called "May Day" so that Operations would have a fix in the event of his having to bail out or ditch his aircraft into the sea, after calling for sometime he received an answer on button "B", he was instructed to call on button "C", the name given was not received owing to poor R/T reception. He received no answer on button "C" so again tried button "B" but this time received no answer. About this time Yellow 2 sighted the coast, his engine by this time was cutting badly, his radiator temperature was 130 degrees and his oil temperature was near 100 degrees and oil pressure very low. On crossing the coast near Start Point, Yellow 2 began to look for a field in which to land, his motor however cut completely and he made a wheels up crash landing in a small field about one mile north of a town which later proved to be Kingsbridge, Devonshire.[39]

Were the Me109s on a weather reconnaissance or a freelance sortie? It is unknown, but in any case the encounter seemed to be pure chance, and luckily the Squadron escaped without a loss. For the rest of February, 402 remained on a state of readiness, engaging in practice flights when conditions permitted.

On 23 February 1942, one of the Squadron's pilots failed to return from a local flight. Sgt Irwin James Eady lost his life when he hit a high-tension pylon and crash-landed near Bournemouth, Dorset. Eady, a 20-year-old from Foresters Falls Ontario, is buried in the Brookwood Military Cemetery, Woking, Surrey.[40]

Three days after Eady's death, four bombed-up Hurricanes took off for RAF Station Ibsley, north of Bournemouth. After landing and refuelling at the southern coast aerodrome, they hooked up with their Spitfire escort from Nos. 234 and 501 Squadrons. With F/L N.H. Bretz leading, they set out in a futile search for shipping reported off the Cherbourg Peninsula. Then, on 28 February, two sections bombed the enemy during a shipping reconnaissance. This operation marked 402's last action flying the Hurribombers. Changes were once again afoot for the Winnipeg Bears.

F/L NORM BRETZ, DFC, FROM WINNIPEG

Chapter 2

March to July 1942

The Squadron bade their Hurricanes goodbye on 4 March, and made their way to RAF Station Colerne, Wiltshire, to begin conversion training on the Supermarine Spitfire VB.

CLIMBING INTO THE COCKPIT IS S/L BOB MORROW - WHILE HIS HURRICANE IS BEING 'BOMBED-UP' AT WARMWELL

One of the best-known and most beloved fighters in history, the Spitfire owed much of its legend to a structure which lent itself to continual improvements. Thus the Spitfire took its place as one of the few types to serve as a front line aircraft from the start to the finish of the war. 402 kept up with the Spitfire improvements during the remaining war years, operating at various periods: the Mk V, Mk IX, Mk XIV and Mk XVI. The switch in fighters happened to be timely for the Squadron because the RAF was preparing for a new offensive at the end of March.

After spending a couple of weeks at Colerne, 402 moved to RAF Station Fairwood Common, South Wales, to undertake Spitfire conversion training. For the most part, this training consisted of defensive patrols and trips to the gunnery range. The Squadron suffered a loss on 29 March 1942, when one of two pilots engaged in a convoy patrol, encountered engine trouble:

FSGTS RELAX ON A SPITFIRE. NOTE THE NAME "JO" ON THE COCKPIT DOOR JUST ABOVE THE CENTRE PILOT'S HEAD

Red Section, Spitfire VB's of No.402 Squadron, RCAF, left Fairwood Common at 1313 hours to patrol the convoy "YOUNG" and were reported in position at 3000 ft at 1336 hours. At 1355 hours Red 2 (FSgt Elliott) called up Red 1 and reported his engine was cutting out, he then repeated this message to Operations who obtained a fix on his position.

Red 2 then turned and glided towards the convoy trying all the time to re-start his engine, Red 1 who was following Red 2 in his glide told him to bail out but by this time he had succeeded in getting his engine to fire a few times and this no doubt encouraged him to keep on trying, finally however he called up to say that he was going to crash land in the sea. Red 2's position on landing was 12 miles south of Tenby and about one half mile from the middle of the convoy. Red 1 saw the aircraft pancake onto the sea, bounce a short distance and turn up on its nose, the aircraft stayed in this position for about 45 seconds before sinking. Red 1 did not see Red 2 in the water after the landing but did see a small unidentified object floating in the patch of oil left by Red 2's aircraft. One ship flying a balloon and two escort vessels went to the spot immediately and one of the escort vessels launched a small boat which made a careful search of the oil patch left by Red 2's aircraft. Red 1 who was orbiting the spot the whole time saw no one picked up by the rescue craft, which after its search returned to the escort vessel, the three ships then returned to the convoy. Red 1 continued to search the area of the crash for some time but saw no sign of Red 2, he then returned to the convoy and continued his patrol until ordered back to base. Further searches were carried out by Squadron Leader R.E. Morrow and Air Sea Rescue boats but these revealed no sign of Red 2.

Lloyd George Elliot was a 21-year-old from Winnipeg, Manitoba. Because his body was never found, and he has no known grave, his name is inscribed on the Runnymede Memorial.[41]

Due to weather and restrictions placed on the offensive, the overall scene was very much at a standstill for the first months of 1942 – at least until British Intelligence learned that the Luftwaffe had redistributed their France-based forces to Norway and Denmark.

WALL ART IN THE 402 SQUADRON LOUNGE, FEATURING P/O PRUNE. PERCY PRUNE WAS AN EGG-HEADED, FICTIONAL SPITFIRE PILOT WHOSE IMAGE WAS POPULAR IN RAF/RCAF UNITS DURING THE WAR. HE WAS THE CREATION OF W.J. HOOPER OF 54 SQUADRON RAF.

The redeployment allowed them to focus on the northbound convoys heading to Russia. To take advantage of the thinning German airpower in France, and to keep the Luftwaffe from recovering, Fighter Command held a meeting in early March to discuss this matter. "On 13 March the Air Staff, therefore, decided to recommence daylight Circus operations over France, the primary aim being to hold and destroy as much as possible of the enemy's fighter strength on the Western Front."[42] The new offensive began on the 24 March with a large-scale Circus operation attacking the Comines Power Station and the marshalling yards at Abbeville.[43]

Orland and John Brown Collection

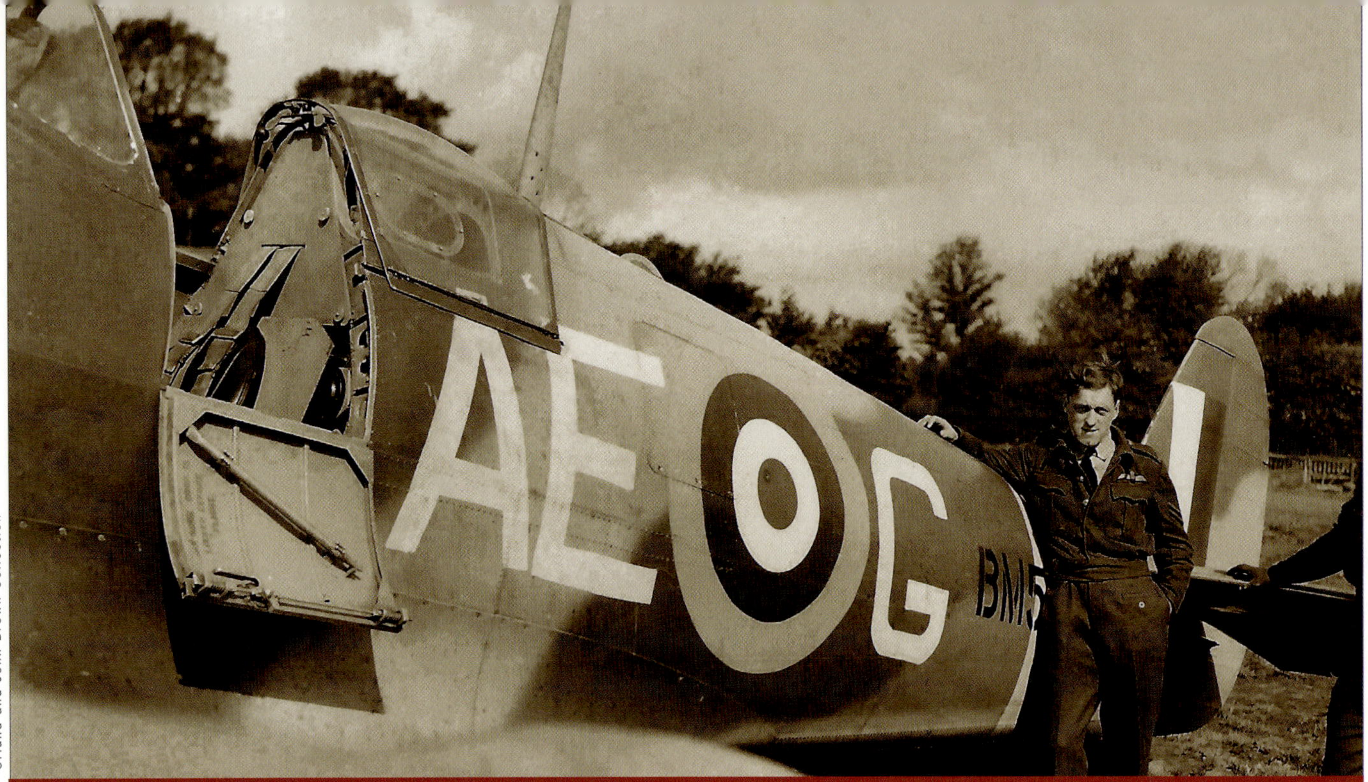

FSGT HARLAND FULLER BESIDE HIS SPITFIRE MK VB

To provide sufficient force for the new offensive, No.11 Group was reinforced with one Spitfire squadron in each sector. At the end of May, as one of the designated squadrons, 402 flew their Spitfires to a new home in Redhill, just south of London. As an integral piece of the new reinforcement, the Squadron again relocated to RAF Station Fairwood Common on 18 March 1942, as an interim guest.

DISPERSAL AT REDHILL

402 Association Archives

The Squadron spent the rest of March training, while most of April involved with defensive patrols. On 25 April, they provided rear cover for six RAF Douglas Bostons (a medium twin engine bomber) on an operation to Cherbourg. This trip proved to be uneventful, but later that same day; they participated in a sweep over the Cherbourg Peninsula. The action resulted in P/O Magee and S/L Morrow each attacking an Me109 without results. The fight initiated the start of a busy summer for the Winnipeg Bears, as Super Circuses[44] became part of Fighter Command's ever-growing repertoire.

The Air Ministry estimated that a total of 200 German day-fighter casualties per month, from all causes on the

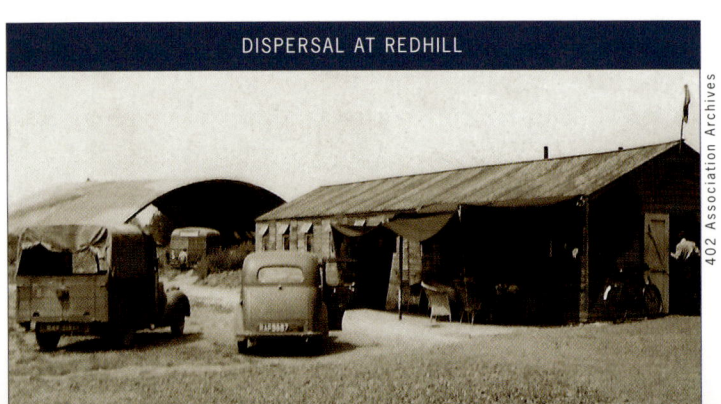

Western Front, would result in declining enemy strength. Also, that 250 fighter casualties would lead to Luftwaffe reinforcements in the West at the expense of fighters being used in Russia. In order to meet the 250 a month mark, Fighter Command had to inflict half these losses in battle. On 24 April the Air Officer Commanding-in-Chief instructed his Group Commanders to inaugurate Super Circus operations, meaning an average of six bomber sorties a day by the Boston and Blenheim squadrons of No.2 Group. Offensive operations in No.11 Group were to be maintained at their current level until the opening of the German spring offensive in Russia, which the Intelligence Branch estimated would take place before the end of May. After this, they aimed to increase the scale of operations by 50%.[45]

Three sections were scrambled in the afternoon of 23 May to join in an air-sea rescue search. As the operation progressed, a 91 Squadron Spitfire flown by Free French pilot Jean-Marie Maridor, mistook 402's Spitfires for enemy fighters, and successfully attacked future 402 O.C., F/L D.G. Malloy, DFC. With his engine on fire, Malloy was forced to bail out over the Channel. The unfortunate airman ended up in Dover hospital for treatment of facial injuries. Seeing red, Malloy's wingman, P/O I.G. Keltie, pursued his leader's assailant and fired upon him. Keltie followed Maridor to Hawkinge where the French pilot crash-landed, causing considerable damage to his Spitfire. Keltie also landed at Hawkinge, but it is unclear what came of the meeting with the pilot or the O.C. of the squadron.

Later that evening, S/L Morrow attempted to fly to Hawkinge, but low visibility kept him from making the trip. Morrow attended a court of inquiry several days later.[46] (Maridor, who obviously knew what he was doing behind the guns of a Spitfire, but needed more training on air-craft recognition, went on to be one of the leading V-1 aces of the war.)

S/L "BUD" MALLOY, LEFT, WITH HIS ADJUTANT

BRUCE INNES WITH HIS SPITFIRE MK VB

By the beginning of June, it had become obvious that the fighter offensive did not justify the hopes, or even the expectations placed on it two months prior. Fighter

SGT J.C. HUGHES ALONGSIDE HIS SPITFIRE MK VB

Command's considerable losses were greater than those inflicted on the enemy. There were several reasons for this, the most important being the marked superiority of the Luftwaffe's newest fighter, the Focke-Wulf 190 (Fw190).

Nicknamed the "Butcher Bird", the Fw190 justifiably ranks as one of the finest fighters of the war, and in the hands of an *experten* (German ace) proved to be a deadly foe. The new German fighter had a better rate of climb, more speed, showed almost as much manoeuvrability as the Spitfire VB (the aircraft which equipped the majority of Fighter Command's squadrons) and it packed a hefty punch with its four 20-mm cannons and two machine guns.

Not only was the new Focke-Wulf giving Fighter Command headaches, the Luftwaffe had also adopted new interception techniques to suit the mass formations coming from Britain. The enemy had also greatly improved his control and warning system since the early spring and had adopted new tactics. Instead of climbing out to intercept RAF raids before they reached the coast, enemy fighters were content to gain height in back areas and then move to a superior tactical position – i.e. up sun and with superior height – from which they could intercept without heavy losses.[47]

Fighter and Bomber Commands needed to change tactics. Instead of penetrating as far as the escort fighters would take them, they were forced to choose targets closer to the coast. In August, the mass Rhubarb[48] operations that 402 and the rest of Fighter Command had been taking part in, were scaled back. They were not achieving the hoped for results, and were too costly through the loss of pilots.[49]

Many of the RCAF and RAF's best fighter pilots were lost on Rhubarbs; including the great Battle of Britain aces W/C R.S. Tuck, DSO, DFC and Two Bars, W/C B. Finucane, DSO, DFC and Two Bars and Canadian, F/L W.L. McKnight, DFC and Bar.

August to December 1942

402 began conversion to the new Mk IX Spitfire on 2 August. The Spitfire Mk IX, one of the major variants of the type, was pressed into action as a stopgap measure to counteract the Fw190. The Mk IX, essentially a Mk V with revised bearers to take the new engine with its four blade propeller and modified radiators under the wings, restored the balance of power.[50] Fighter Command had in the Mk IX, not only an aircraft that could compete with the Fw190, but the Spitfire variant of preference for many successful exponents of the famous fighter. 402's conversion took place during the days prior to relocation to RAF Station Kenley, Surrey, a short distance away.

Along with the arrival of the new Spitfires came a new O.C., S/L N.H. Bretz who assumed command on 17 August. Norman Bretz had been one of the originals of No.2 Squadron's "B" Flight back when the unit converted from Lysanders to fighters on 11 December 1940. He was one of the original British Commonwealth Air Training Plan

Art by Pat McNorgan

SPITFIRE MK IX COLOUR PROFILE

THE ORIGINAL NO.2 SQUADRON PILOTS, DECEMBER 1940

graduates (BCATP) and later, helped define the standard procedure of circling a downed airman in the water.[51]

The change to the Mk IX Spitfire proved timely, providing the Winnipeg Bears with an aircraft that offered a more effective aerial umbrella during the famous Dieppe raid. The Squadron took an active part in the air battle over Dieppe, performing four operational flights in direct support of the 19 August raid, named Operation Jubilee. The first sortie saw 402 carry out a high cover patrol of Dieppe. During the second, they escorted 24 Boeing B-17 Flying Fortresses to Abbeville. On the third sortie, involving a sweep over Dieppe, they engaged the enemy and damaged three Fw190s. For their last 'uneventful' outing, 402 flew a high altitude patrol over the bloody French beach. "The four sweeps were carried out without loss to the squadron. The entire groundcrew also did a fine job of work continually throughout the day, not leaving the dispersal point from dawn until well after sunset."[52] The following day, as part of the Kenley Wing, the Squadron again provided escort for B-17 Flying Fortresses bombing Amiens. From this point until the end of the year, the unit operated in the thick of the action.

402 SPITFIRE MK IX BS 306

402 suffered a double loss on 24 August when they bounced eight Fw190s during an afternoon Circus to Le Trait, France. The hunter turned into the hunted when the Focke-Wulfs proved to be a decoy, and another forty Fw190s from Jagdgeschwader JG 26 pounced on the Canadians.[53] During the ensuing dogfight, P/O G.P. MacKay and FSgt V.H. Miller were shot down. F/L E.A. Bland and P/O I.G. Keltie were badly wounded but survived. 23-year-old Gerald Peabody MacKay, from Rock Island, Quebec and 21-year-old Victor Howard Miller of Richmond, Indiana, U.S.A., rest side by side in the Franco-British War Cemetery, St. Valery-en-Caux, near Dieppe, France.[54]

The nuisance "tip and run" raiders were becoming a thorn in Fighter Command's side. The Luftwaffe typically used the Fw190 for these operations. The fighters concentrated mainly on the south coast, coming low and fast over the Channel in order to stay under the radar screen. Once reaching the coast, the raider would perform a rapid climb to gain altitude. The tactic involved

JERRY MACKAY, KILLED IN ACTION 23 SEPTEMBER 1942

F/L E.A. BLAND SITTING ON THE COCKPIT DOOR OF HIS SPITFIRE "MARY ANNE II"

dropping a single bomb on a designated target, then strafing targets of opportunity before quickly pulling out. "Tip and runs" were highly dangerous for the attacker and required great skill to successfully carry out.

S/L D.G. Malloy succeeded S/L Bretz as O.C. on 27 September 1942, and served an eight-month tour as leader of the Winnipeg Bears.

Fighter Command instituted standing patrols towards the end of September to combat these tip and run raiders. This new directive affected the Squadron, involving them in many new patrols along the south coast between Mayfield, Beachy Head, Shoreham, Brighton, Dungeness and New Haven. They carried out this secondary operation into March of the New Year with very little result.[55]

A 402 MK IX SPITFIRE

CF Photo PL-15058

On 8 November, the offensive picked up again due to Operation Torch, the Allied invasion of French North Africa. Churchill chose to assist the opening phase of the operation by intensifying the air offensive. He had hoped this would create a diversion, keeping German fighter squadrons from being deployed to the Mediterranean and also prevent the Luftwaffe units already there from being sent to Africa.[56] So in between their standing patrols, 402 took part in several uneventful escort missions to France, but the weather severely limited their operational flying time. No flying took place from 11-17 and 21-23 November.

December proved to be a busy month for the City of Winnipeg Squadron. They participated in several operations, for the most part providing escort to B-17 Flying Fortresses and B-24 Liberators of the United States 8th Air Force. Very few enemy aircraft were encountered, but on 4 December the Squadron lost two pilots when Sergeants R.B. Honeycombe and H.E. McGraw failed to return. 20-year-old Richard Byrne Honeycombe, from Brooklyn, New York, U.S.A. is buried in the Souvenir Cemetery, Longuenesse, France, and Hugh Edward McGraw, from Kirkland Lake, Ontario, is buried at the War Cemetery at Pihen-les-Guines, Pas-de-Calais, France.[57]

January to December 1943

Shortly after the New Year, 402's operations increased as the pendulum of air superiority swung in favour of the Allies.

On 17 January, the Squadron became involved in a furious dogfight over the Bolbec area near Le Havre, France. During the second Rodeo of the day, 402, operating with 401 RCAF and 412 RCAF squadrons, formed a wing consisting of 24 Spitfire Mk IXs and 15 Spitfire Mk VBs. Crossing Beachy Head on the deck, the Wing climbed to 9,000 feet at Veulew Les Roses and proceeded to Bolbec. Here, under the command of W/C J.C. Fee, DFC and Bar, 402 and 401 Squadrons attacked a series of locomotives while 412 Squadron provided cover. On the climb to rejoin their flight after the last strafing run, Yellow 1 and 2 received a warning over the R/T that nine Fw190s were attacking in a shallow dive from cloud cover at 8,000 feet.

Yellow 1 and 2 turned toward the German fighters in a head on attack, while the rest of the flight and 412 Squadron joined the developing dogfight. Four Focke-Wulfs engaged Yellow 1 and 2 on the climb, but the Canadians inflicted damage on two of the enemy aircraft. Even though the flight became separated, Yellow 3 and 4 damaged two more Fw190s. Red 3 and 4 broke away from the main battle to pursue a Focke-Wulf, but as Red 3 shot it down, he found himself without his wingman, P/O A.M. Skinner. In the meantime Red 1 and 2, W/C Fee and his wingman F/O M.J. Sunstrum, were also in trouble. At 1355, the Wing Commander transmitted his last message, "I'm hit fellows and I'm going out." Both Fee and Sunstrum were never seen again. Skinner also died that day and although no one saw him go down, his body was recovered.

P/O ALLISTER MACLEAN SKINNER

An estimated 20 Fw190s engaged the three Spitfire squadrons, with the Wing claiming one destroyed, one probable and five probably damaged. But the price was dear. 402 lost two pilots in addition to W/C John Clarke Fee, DFC and Bar, from St. Williams, Ontario. Because the 23-year-old Fee and 21-year-old Michael Joseph Sunstrum of Naicam, Saskatchewan have no known graves, their names are inscribed on the Runnymede War Memorial, Englefield Green, Egham, Surrey. Allister MacLean Skinner, age 25, from Stellarton, Nova Scotia, rests at the War Cemetery at Grandcourt, France.[58]

402 continued to spend more time supporting US 8th Air Force bombers as the Americans gained operational experience with their daylight bombing raids.

The unit provided escort and carried out diversionary sweeps to confuse the German radar. During one such mission, the Squadron shared in a significant milestone. On the 27th they participated in a Roadstead[59] to Dunkirk where enemy aircraft were encountered. In the ensuing melee, 402's F/O L.M. Cameron (a future postwar commanding officer) along with 403 RCAF Squadron, shared in the destruction of the 500th enemy aircraft of RAF Station Kenley:

THE NEVER ENDING GAME OF CARDS. DISPERSAL HUT, KENLEY 1940 – L TO R: WO ORLAN BROWN, P/O GEORGE KIETH, P/O SUNNY SUNSTRUM, FSGT AL SKINNER, P/O LORNE CAMERON

This Sector destroyed its 500th enemy aircraft in yesterday's operation. Upon close examination of the Combat reports, it is impossible to ascertain which of the three aircraft was shot down first. The honour, therefore, of destroying this 500th Hun is to be shared between Squadron Leader L.S. Ford and Pilot Officer E.L. Gimbel of No.403 Squadron and Pilot Officer Cameron (picture) of No.402 Squadron. Signed H.L.B. Hodson, Wing Command Flying, RAF Station Kenley.[60]

F/O Cameron's combat report adds further detail:

I was flying as Blue 3, Hoboe Squadron in the mid-channel area north of Dunkirk and was watching four F.W. 190's [sic] follow us out, with more behind them.

When the Fw.190's [sic] started to come in close behind us the C.O. gave the order to break. Blue 4 and I broke to port into the outside enemy aircraft of the four. This engagement took place at 23,000 feet. This Fw.190 was coming in from two o'clock and 1,000 feet above. We pulled straight up into him closing, he broke to starboard leaving us right on his tail. I closed to 100 yards and gave him a four second burst with slight deflection and could see strikes along the port side of his fuselage near the wing roots. He slowly turned over and dived away. I continued to give him another five second burst and he started to spin down. I followed him to 17,000 feet and broke away as more Fw.190's [sic] were above us, and our friends were being engaged. Seeing it was no use to climb back again I went right down to the deck. After I had pushed the nose straight down I saw a splash in the water just to my right which was about 6 or 7 seconds after I had finished firing.

When I was about 3 or 4 thousand feet and coming straight down for the deck I saw two other bright white rings where aircraft had hit water. I proceeded to base on the deck. I claim one Fw.190 as destroyed.

The action was witnessed by F/O N.A. Keene, Blue 4, who backed up Cameron's claim.

Flying as Blue 4, Hoboe Squadron made the break at 23,000 feet along with Blue 3 and carried out the attack with him: "I fired a short burst of M.G. as my cannons did not work. I estimated my range to be 200 yards and I was about 100 yards being Blue 3. I saw Blue 3 hitting strikes on the Fw.190 along the left side of the fuselage. The Fw.190 sort of mushed into the air and began a slow spin, at this time I observed the splash of an a/c hitting the water. I could not continue firing because Blue 3 was in the way but I covered him down to his break off at approximately 17,000 feet. The Fw.190 was still spinning towards the sea when I last saw it. This was a second after Blue 3 had broken off. I followed Blue 3 out on his Jinking Dive towards the deck and lost sight him as he entered the haze.

At this, probably at 7 or 8 thousand feet, I saw one splash in the sea to our right and a split second later saw two more splashes in the sea further over to our right.

The Squadron took part in many Circuses, Rodeos and Sweeps before being relocated to RCAF Station Digby, Lincs, in No.12 Group on the 21 March. Here they became part of another famous fighting team, the Digby Wing, and underwent training to combat the E-boat menace. As the Squadron travelled north by train, they left behind their beloved Spitfire Mk IXs, which were badly needed by No.11 Group.

BRUCE INNES PREPARING FOR FLIGHT IN FRONT OF S/L NORTHCOTT'S SPITFIRE EP 120

Spitfire Vcs were waiting for the Squadron at Digby. Long-range fuel tanks marked the difference between these Mark Vcs and the ones 402 had previously used in 1942.

The added fuel tanks increased the radius of the Mk Vc, enabling it to range as far as The Hague in Holland or even half way to Paris.

No operational flying took place the rest of the month, save for a section scramble on 23 March. Much of the Squadron's time up until mid-May involved an intensive training schedule which included: trips to the gunnery range, formation flying, affiliation flights with bombers and practice landings on a dummy deck. Operational flying increased in April with a diversion sweep to Ostend, Brussels on 4 April, a convoy patrol on the 8th, and several Air Sea Rescue flights.

On 14 May the Winnipeg Bears successfully acted as rear cover for a bomber attack on shipping off the Dutch Coast. The next day, a former member of the unit, S/L L.V. Chadburn, assumed command of 402. After promotion to Wing Commander a month later, Chadburn became widely regarded as one of the finest Wing Leaders of the war. The U.S. 8th Air Force knew him as "The Angel" and his squadrons were "Angel Squadrons" because of the outstanding escort work provided for the American heavy bombers. His death in a mid-air collision with another Spitfire shortly after D-Day devastated the Squadron. Lloyd Vernon Chadburn, age 24, of Aurora, Ontario, is buried in the Ranville War Cemetery, Calvados, France.61

NORMAN BRETZ (LEFT) WITH LLOYD CHADBURN

SUPER GROUNDCREW: L-R BOB BIRT, BENNY DOMINI, UNKNOWN

The Squadron provided escort for RAF Mitchell bombers destined for Flushing on 31 May and got into the thick of it when six Fw190s jumped them. Chadburn led the Squadron into a sharp break, but Blue Four, F/O J.G. Torney, took fire, which forced him to bale out. S/L Chadburn dived after one the Focke-Wulfs. Although his rounds hit home, he could only claim it as a "probable." This encounter marked the beginning of a busy 12 months for 402.

During this period, the Winnipeg Bears saw plenty of action, mostly from anti-shipping strikes along the Belgian, Dutch and northern French coasts. The Squadron's redeployment to RAF Stations, Merston, 8 August to 19 September 1943 and Wellingore, 9 February to 21 March 1944 were the only interruptions during this busy twelve month period.

THE SQUADRON'S "PUKKA GEN" BULLETIN BOARD

At this time Fighter Command instituted a new type of operation referred to as a Jim Crow,62 or a combined weather

(FSGT K.W.) "BIRD'S ARRANGEMENT OF SPITFIRE" – AS NOTED IN THE PERSONAL PHOTO ALBUM OF ORLAN BROWN – BIRD WAS UNINJURED IN THE CRASH OF "EL LIBERATOR" MK VB BM 246

and a shipping reconnaissance operation. Four Squadron pilots carried out this new task on 25 May, encountering flak from R-boats. These so-called *"Räumboote"* or R-Boats were small minesweepers designed to operate in shallow waters such as harbours, coastal areas or rivers.[63] 402 took part in a pair of Jim Crows over the next few days with shipping encountered on both trips.

This sudden surge of activity also coincided with Operation Point Blank, a strategic air offensive against the German aeronautical industry. Point Blank opened with U.S. 8th Air Force attacks on Halberstadt, Brunswick and Oschersleben on 12 January 1944, and became the American heavy bomber's main focus until the middle of 1944.[64] Diversionary sweeps and attacks were carried out by the medium bombers and Fighter Command in the hopes it would distract and confuse German radar when the heavies crossed into Germany unescorted. When not engaged in these missions, the Squadron assisted in providing support to medium bombers attacking E-boat bases at Ijmuiden, Ostend, Boulogne and Cherbourg.

Twelve aircraft, under 402's new skipper, S/L P.L.I. Archer, DFC, flew to a forward airfield, RAF Station Coltishall, located north of Norwich on the east coast, where they escorted RAF Beaufighters on a successful but costly shipping strike on 17 June. Chadburn had relinquished command to S/L Archer on 13 June 1943. Archer had seen action and been wounded while flying Spitfires with 92 RAF and 416 RCAF Squadrons. The six-victory ace had been slated to take command of 421 Squadron the very next day but did not return from the Beaufighter escort sweep. His four

CHAPTER 2

THE ACE FROM THE BARBADOS, S/L PHILLIP ARCHER BY HIS SPITFIRE – NOTE THE "ARCHER" NOSE ART

days in command mark the shortest tenure of any O.C. in the Squadron's history. Phillip Leslie I. Archer, 27 years of age, from Hasting St. Michael, Barbados, rests in St. Omer Cemetery, Longuenesse, France.[65]

S/L JEFF NORTHCOTT STRAPS ON HIS CHUTE FOR ANOTHER OP

On 18 June, S/L G.W. Northcott, DSO, DFC and Bar, took command of 402 after the death of S/L Archer. 402's CO of four days earlier, S/L L.V. Chadburn DSO, DFC and Bar, was promoted to W/C (flying) of the Digby Wing. The team of Northcott and Chadburn displayed an easy, confident and inspirational leadership style that took a back seat to no one. Northcott's tenure of command lasted

THE GREAT 402 TEAM OF S/L GEOFFREY "JEFF" NORTHCOTT (ON WING) AND W/C "CHAD" CHADBURN

for over a year, making him the Squadron's longest serving wartime O.C. As an example of Northcott's attention to the morale of the entire Squadron, he initiated a new routine. Whenever the fighters returned from an operation, the pilots would form tight formations of three sections in line astern a short distance from the aerodrome. Northcott would then lead his men on a low-level, high-speed pass over the station, giving the groundcrew an inspirational display. After the pass, they would break up and outward into three units and in an orderly and disciplined manner, and follow each

WE STAND ON GUARD

S/L NORTHCOTT (SECOND FROM RIGHT) IS BRIEFED ON AN ENGINE SNAG

other into the landing pattern. It became a matter of great pride that the Winnipeg Bears put on a good show.

As the year ended, the Germans were readying the V-2, the next generation of pilotless bombs to be launched against Britain. Far deadlier than the V-1, or "Doodlebug", (still yet to be launched) the V-2 was a product of Wernher von Braun's team of rocket scientists at Peenemunde. It was capable of delivering a 2,000-pound payload to a target up to 230 miles away. About 2,500 rockets were launched by Germany with about 500 of them aimed toward London. Flying at supersonic speeds, the V-2 was virtually impossible to bring down and its launch sites became a priority for the medium bombers.

About the same time, the single-engine fighter force, mainly the Spitfires, were being equipped with bomb racks enabling them to carry one 250- or 500-lb bomb. As part of the RAF's pre-invasion plans, these bombed-up Spitfires allowed the squadrons to both operate in a tactical manner, and fill the short supply of tactical aircraft.

402'S ACE S/L JACK MITCHNER, DFC – PHOTOGRAPHED WHILE O.C. OF 416 SQN

January to October 1944

At the beginning of 1944, the Squadron continued to be engaged in the offensive, providing escort to medium bombers and to the North Coates, anti-shipping Beaufighter Wing.

On 4 February, the Winnipeg Bears began two months of intensive high- and low-level dive-bombing training. With the invasion of the continent on the horizon, fighter-bombers were urgently required.

In mid-February, the Squadron relocated to RAF Station Wellingore, Yorkshire, (a satellite of Digby) which became their new home for the next month and a half. Eleven aircraft were deployed to RAF Station Catterick, Yorkshire, on 16 February to participate in Exercise Eagle. One of the pre-invasion exercises, this operation tested the lines of communications between the Army and Air Force during simulated combat conditions. The Squadron tasked eight aircraft to fly at low level, locate ammunition dumps and beat up concentrations of troops, tanks and artillery. This exercise proved quite successful, with 402 airmen able to locate their targets on each occasion. The Digby Wing received credit for holding up the enemy advance by one day. The Squadron returned to RAF Station Digby on 24 February after carrying out 89 low-level attacks. Unfortunately during this move, an incident occurred involving two Spitfires and resulting in one fatality. The following was recorded in 402's Operations Record Book (ORB):

THE BOYS HAVING SOME FUN IN THE SNOW, WELLINGORE, FEBRUARY 1944 – L TO R: DEAKIN (SITTING), ROACH, BASS, MCLAREN, HOLMES

The Squadron was moving over to Digby from Wellingore today, and F/O D. Sherk with F/O Morris was detailed to proceed ahead and take over

FROM THE RIGHT: W/C LLOYD CHADBURN, S/L JEFF NORTHCOTT, F/L JACK MITCHNER

readiness at Digby. They took off in formation with F/O Morris, Blue 1, and F/O Sherk, Blue 2. They circled the field and came in fairly low over the dispersal hut.

Blue 1 then started to do a slow roll towards Blue 2. Blue 2 did not realize what the manoeuver [sic] was until too late and while he did all he possible could to avoid it, Blue 1 hit the starboard wing of his aircraft twice. Blue 2's aileron jammed, and controls, and he was forced over onto his back. Blue 1 spun into the ground and was instantly killed. Blue 2 after considerable difficulty righted his aircraft and regained control. One wing was slightly buckled but he made a very skillful landing at Digby.

25-year-old Arthur John Morris of Winnipeg, Manitoba is buried in the Scopwick Church burial ground, Scopwick, Lincolnshire, England.[66]

Poor weather put an end to all flying for the rest of the month. When it finally cleared on 1 March, eleven aircraft from 402 joined with 234 Squadron to take part in a Roadstead to Den Helder, in the Province of North Holland, Netherlands. The Squadron escorted twenty-two RAF Beaufighters from 143, 236 and 254 Squadrons (North Coates Wing), which sunk the Dutch merchant vessel "Maasburg".[67]

The Squadron's ORB describes the attack:

> [Rendezvous] R/V made as planned and operation successfully carried out. Only one ship 3/4000 tons seen off Den Helder. Reported escort vessel no longer with it. Vessel appeared to be on sand bank and was attacked with cannon and torpedo. Numerous cannon strikes seen and two torpedo hits, one amidships and one on bows seen to hit, both on starboard side.
>
> No enemy aircraft but moderate light flak from three guns on ship, was experienced. Our aircraft orbited ship while some Beau's made two more runs.

Several more Lagoons[68] were carried out before the Squadron relocated back to Digby on 21 March. 402, along with 64 Squadron, again escorted 45 Beaufighters ten days later for an uneventful Roadstead to Holland. Operational flights did not occur with great frequency during this period. The weather cancelled shows, and with more time spent on training flights, the monotony was soon wearing on the pilots. In fact, the Squadron's diarist thought it worth men-

MOULDIE, A 402 MASCOT, WAITS FOR A RIDE BACK TO BARRACKS – DIGBY

tioning in the entry on 4 April. "Weather still closed in and all flying activities remain at a standstill. Pilots are becoming 'fed up' with the inactivity."[69]

Gearing up for the Normandy Landings soon provided 402 with the action they were missing. On 12 April, the Squadron deployed to No.15 A.P.C., RAF Station, Peterhead, Scotland for a 10-day air firing and low-level attack course. 402 excelled on the course, setting a new record for the number of sorties carried out in a single day.[70]

By the end of April the Squadron, under the command of New Zealander W/C J.M. Checketts, DSO, DFC, had relocated to an Advanced Landing Ground (A.L.G.) located at

RAF Horne where they became part of No.85 (Base) Group, No.142 (RAF) Wing, Air Defence of Great Britain (A.D.G.B.).[71]

F/L "DOC" GATES, 402'S MEDICAL OFFICER SHOWS THE PROPER TECHNIQUE OF SETTING UP A TENT

The new field assisted the preparation for cross-channel operations when the Second Tactical Air Force (2nd TAF) was formed. Primarily 2nd TAF assisted in the preparatory work for the upcoming invasion. Until then, providing escort to the medium bombers and generally assisting readiness for the invasion became the responsibility of A.D.G.B. Requirements for the protection of Britain, marshalling areas and shipping lanes fell upon the newly formed No.85 Group, A.D.G.B. The work fell to six squadrons, including 402.

As D-Day grew closer, the pace of escort missions for medium bombers attacking the invasion area and other parts of France increased, keeping the Squadron quite busy during the month of May. Many of these operations were uneventful because, since the summer of 1943, the Luftwaffe had been slowly withdrawing their fighter force from France.[72] When not engaged in escort missions or sweeps, the Winnipeg Bears took part in standing patrols, instituted at the end of April by A.D.G.B. "From April 20th until D-Day A.D.G.B. maintained daily standing Spitfire patrols ranging in strength from 6 to 40 over the vulnerable assembly areas of the British and American invasion fleets such as the Isle of Wight, Lyme Bay, Portland, Falmouth and the Lizard (Point) area."[73]

P/O PRUNE ART ON 402 MAE WEST

CRASH OF A SPITFIRE MK IX BELIEVED TO HAVE BEEN PILOTED BY F/L W.G. DODD, DFC

The quick operational tempo carried over into June as the Allied invasion of Europe drew closer. Dusk patrols on 2 June and close escort to RAF Beaufighters on the 3rd, were uneventful. The day before the invasion, the Squadron placed one section on a state of readiness from sunrise to sunset, while the rest of 402 carried out 48 patrols over the Solent, Beachy Head and Foreness Point areas.

On 5 June, the pilots were briefed about the momentous event planned for the next day. During the briefing they learned about their role in the opening phase of the D-Day landings.

INVASION STRIPES ON A SPITFIRE MK VB

On the eve of what became known as "the Longest Day", black and white identification or 'Invasion Stripes', were painted on the upper and lower wings and aft fuselage of all Allied aircraft taking part.

402's job entailed low-level patrols over the beachhead, and they completed three uneventful patrols. Three more were carried out on D +1, (the day following D-Day) again without incident.

Weather forced the Squadron to scale back to two patrols on 8 June, and no patrols on the 9th. When it cleared the following day, four patrols were mounted.

During one, P/O C.H. Bavis was shot down over France while attacking a gun post. He successfully bailed out and landed in Allied territory. Because of the amount of traffic plying back and forth between France and Britain, Bavis was able to hitch a ride with a Royal Navy Motor Torpedo Boat. Once back in the U.K. he caught a train and returned to base at 1700 hours.[74]

On the evening of the 13th, six pilots landed in France on an Advanced Landing Ground (ALG) after completing their patrol. These Advanced Landing Grounds were opening up on the beachhead and were made possible by the hard work of construction crews toiling around the clock.[75]

For the first time since May 1943, Bomber Command began to operate during daylight hours on 14 June. The "City of Winnipeg Squadron" assisted the escort of 153 Lancasters and 6 Mosquitoes of Nos.1, 3 and 8 Groups, whose target was an important railway marshalling yard at Chartres on the outskirts of Paris. Unfortunately, because of cloud over the target, only 12 Lancasters were able to bomb the yards.

A German V-1 pilotless flying bomb screeched over 402's station the night of 15 June. The flak gunners put up an intense barrage and some explosions from the V-1 could be plainly heard. The launching of the first 'Robot' planes fired at England took place during the evening of the 13th and 14th. Thus began a steady stream of these revenge weapons, which appeared in the skies over southern

THE V-1 "DOODLEBUG" FLYING BOMB, WHICH OCCUPIED 402 SQUADRON DURING THE LATE SUMMER AND EARLY AUTUMN OF 1944

England until the launching sites were over run in late summer. The last of the V-1s hit Orpington in Kent on 27 March 1945.[76]

To combat this new threat, anti-aircraft defences at the station were reinforced with twin-Browning .303 machine-guns and any qualified individual was allowed to 'have a go' should the opportunity arise. One such occasion presented itself during the evening of 18 June. A V-1 came streaking over the base and a pilot near a gun pit got in a burst at the low flying "Doodlebug" which crashed a mile away without exploding. Upon examination, it was found to be riddled by .303 light arms fire. Whether this individual belonged to the Squadron or Station remains a mystery.[77]

The unit relocated to RAF Station Westhampnett, in West Sussex along the south coast of England on 19 June. They flew from this base for nine days before moving to nearby RAF Station Merston. This switch brought them closer to Portsmouth, the main area of operations for the navy and sea traffic going to the Normandy coast. 402 carried out patrols of the shipping lanes and the beachhead. The change of location, probably promoted by the Allies establishing a foothold on the beachhead, enabled the A.D.G.B. squadrons to adopt a more aggressive attitude.

The 22nd of June dawned fair and warm, but it heralded a tragic day for the unit. During the morning patrol just northwest of Forest De Cerisy, France, two 402 Spitfires were shot down and a third damaged. "As the patrol was over our own territory the only conclusion that can be reached was that our own flak did it. The aircraft had passed over the area twice and the third time they were met by a terrific barrage of flak."[78] The third Spitfire, flown by F/O J. A. MacLeod, returned to base before the Cat. B damage was noticed, and although P/O N. P. Murphy survived being downed by friendly fire, F/O K. M. Collins did not. A popular fellow with 402, the loss of Collins keenly affected the Squadron.

The fact that Spitfire Mk Vs had been over the area so often made the mistaken identity harder to accept and heightened the tragedy. Kenneth McRoberts Collins, a 24-year-old from Los Angeles, California, U.S.A., rests in the Bayeux War Cemetery, France.[79]

SPITFIRE WHEELS UP LANDING

The black cloud continued to hang over the Squadron. The day after the loss of Collins, as 402 crossed the Channel from France during the second evening patrol, F/O W.G. O'Hagan's engine packed up. Turning back, in an effort to make the French coast, O'Hagan disappeared from view in the poor visibility. Last known to be twenty miles northeast of Barfleur, France, Walter Gerald O'Hagan, a 21-year-old from Montreal, was never seen again. Since O'Hagan has no known grave, his name is inscribed on the Runnymede War Memorial, Englefield Green, Egham, Surrey, England.[80]

On 6 July, 402 engaged in two Ramrods, marking their return to offensive flights into France. They also received exciting news about re-equipping with Spitfire Mk IX's. "The pilots are very enthusiastic over this good news and are very anxious to see and fly the new kites."[81] The first ten were flown in on the 11th and the following day, the remaining eleven brought the Squadron up to strength. While test flights and fine-tuning were carried out on the Mk IX's, 402 soldiered on with the Mk Vs.

S/L G.W. Northcott relinquished command of the Squadron on 28 July to S/L W.G. Dodd, DFC. Wilbert George Dodd, an ace, had been with the Squadron since May, and brought extensive experience with him, having seen action in Malta.

In addition to the Ramrods, one more escort of RAF Heavies kept 402 busy until their relocation to RAF Station Hawkinge, Kent, located on the coast north of Folkestone.

On 9 August, 402 re-equipped yet again with the latest version of the Spitfire; the Griffon powered Mk XIV. This move came about as A.D.G.B. made adjustments of its day fighter squadrons. Along with 350 Squadron, 402 exchanged their aircraft with 91 or 322 Squadrons, who were going to join 2nd TAF.[82]

SPITFIRE MK XIV COLOUR PROFILE

With the new Spitfires in place, the V-1 flying bomb program became a priority for the Winnipeg Bears. They carried out anti-diver patrols (intercepting V-1s) over the Channel and along the English coast. Trying to shoot down Doodlebugs brought its own special danger. Its one-ton warhead packed a considerable explosion, which, if fired upon and hit at close range, could destroy both hunter and prey. Because of this, pilots developed a new method of bringing the bombs down. This involved flying alongside and flipping the V-1 with a wing, or even allowing the airflow from the top surface of the wing to upset the sensitive gyroscopic system, thus sending the bomb plummeting to earth. The Spitfire XIV, with its increased speed (in excess of 40 mph faster than the Mk IX) was an excellent fighter to combat the V-1. 402's pilots reported their first impression of the potent Mk XIV Spitfire and "are very pleased with performance and easy handling of the new aircraft."[83]

The patrol pace became quite steady, with the interception of V-1s setting precedence for A.D.G.B. However, the Allied advanced through France meant the launching sites were being overrun, thus the opportunities for sighting a "Doodlebug" became fewer.[84]

Nonetheless, the Squadron did enjoy some success against the flying bombs. As of 16 August 1944, three V-1s had been shot down and 42 anti-diver patrols completed. F/O A.H. Vickers downed a Doodlebug, which crashed very close to the Squadron dispersal, breaking glass in the pilot's rest room and giving everyone on base a scare.[85] His report of the action states the following: "Attack on Diver 2000' four miles west of Cape Gris Nez. Closed to ack ack zone off Folkstone using all ammo. Saw strikes and Diver lost 1000', broke off attack, at ack ack zone. Diver speed 380. Diver crashed south of Hawkinge 'drome at approx. 0715 hrs."[86]

Twenty-eight sorties were flown the next day with three V-1 sightings recorded. F/O W.D. Whittaker was hit and slightly injured by friendly flak while chasing a Doodlebug during the early hours of the morning but damage to his aircraft was Cat. AC.[87]

The other two successes against the V-1 included FSgt W.G. Austin's score on 23 August and F/L J.A. De Niverville on the 24th. FSgt Austin's narrative of the action describes how dangerous chasing these missiles could be:

JOSEPH HOCTOR LEFT, AND F/L J.A. DE NIVERVILLE. NOSE ART TO THE LEFT OF HOCTOR IS A FEMALE FIGURE RIDING A V-1 BOMB NAMED "BRUNHILDA"

I first saw the Diver about 2 miles North of Ashford at 3000'. I turned and dived from 6000', getting line astern of it about 2 miles SE of Maidstone. I closed to approximately 200 yds, firing about a 5 second burst with .50 and 20 mm. I observed strikes on starboard wing and the Diver began losing altitude. At this time red rockets were fired in front of me warning of balloons. I broke to starboard and saw a column of black smoke rising from a point about 2 miles on what was then my starboard side. Cine gun used. Bromley Observers' Post confirm this claim.[88]

As the V-1 launching sites were being over-run, their threat diminished with each passing day. Hence the Squadron received word that they were to be used on as many first class operations as possible, the diver patrols becoming a secondary operation. Relief from the tedium and monotony of diver patrols pleased the pilots who, with their new Griffon-powered Spitfires, were anxious to mix it up with the Luftwaffe.

As October progressed, 402, along with 350 (Belgian) Squadron, took part in Ramrods and Sweeps along the French countryside and Dutch coast. Most of these sorties were uneventful, but the Winnipeg Bears did try their hand at shooting up rail targets, locomotives and motor transports with the new Spitfires.

On 1 September, the Squadron escorted 50 Halifax bombers, which were part of an even larger bombing force

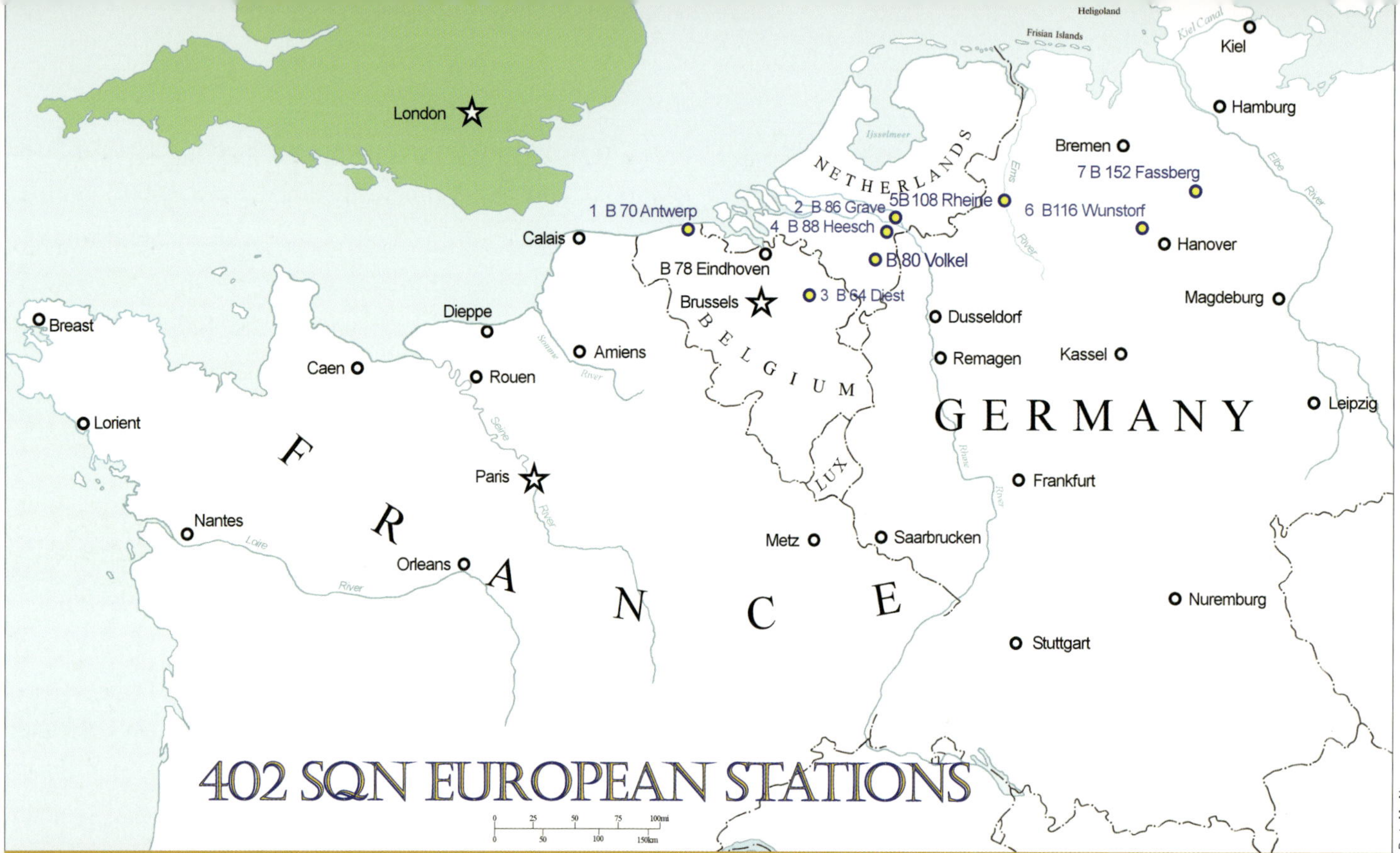

402 SQN EUROPEAN STATIONS

MAP OF 402 STATIONS ON THE CONTINENT. STATIONS ARE NUMBERED IN ORDER FROM WHEN THE SQUADRON FIRST DEPLOYED TO THE CONTINENT

targeting V-2 rocket stores. The operation proceeded as planned although three aircraft, along with one escort, were forced to return to base because of mechanical trouble. After the bombers encountered heavy and accurate flak over the target area, 402 proceeded to launch attacks, damaging locomotives and trains five miles north west of Mans, between Tourai and St. Amand, and also between Lille and Donai heading north. They continued with assaults on a goods train west of Lille at 0930 hours and a stationary locomotive at Orchies, which they left spouting steam following a large explosion. The pilots also observed strikes on the barges they attacked on a canal five miles north west of Roubaix.

A high level meeting held on 13 September discussed the transfer of squadrons between A.D.G.B. and 2nd TAF. A.D.G.B. had begun to plan for a long-range escort group to support the RAF's heavy bombers on daylight raids. A

CHARLES H. "CHUCK" BAVIS AND TOMMY GORDON IN FRONT OF A SPITFIRE MK XIV. NOTE THE SQUADRON'S TIGER MOTH BEHIND THE STARBOARD WING AND THE LONG-RANGE FUEL TANK BETWEEN THE LANDING GEAR

further meeting between the two organizations, held on 25 September, finalized the exchange. The seven squadrons involved, participated in a shuffle that saw 350 and 402, trade places with 132 (RAF) and 441 (RCAF) Squadrons, the latter two already being on the Continent. This came into effect on the 29th with the actual transfer happening on 30 September.[89]

The exchange also had an effect on the servicing echelons (the maintenance units attached to the squadrons). No.6402 would stay behind becoming responsible for the Spitfire IX's of the incoming squadron, while servicing echelon, No.6441, remained on the continent, taking over the maintenance of 402's aircraft. This sped up the transference of the squadrons, but the splitting of the team greatly upset the Winnipeg Bears, particularly the groundcrew, who were not going along with their pilots.

GROUNDCREW AT HAWKINGE

Six Dakota aircraft arrived in the early hours of the 30th to collect personnel and equipment before airlifting them to aerodrome B.70 in Antwerp, Belgium. The Squadron's

SQUADRON GROUNDCREW LEAVING THE SHETLANDS FOR HAWKINGE ON A DAKOTA TRANSPORT

Spitfires followed suit, arriving later that day. Once there, 402 learned they were being moved to airfield B.82 at Grave, Holland, joining No.125 (RAF) Wing. The transfer became effective immediately. After unloading their gear and equipment from the Dakotas, the Squadron spent the night at B.70 before being transported by road to B.82 the next day. Travelling to the new base gave the personnel a first hand glimpse of the ground action results in Belgium and Holland. Upon arrival at Grave, the pilots asked when their groundcrew would turn up. They were experiencing some delay, came the reply; German counter-attacks had cut off the road from Eindhoven. Hours later, the British Guard Brigade had re-opened the road. Descriptions of some of the sights are contained in the Squadron's ORB.

Along the wayside could be seen the damage caused by the advancing allies. As we moved into Holland quite a number of burned out tanks and vehicles could be seen on the roadside. They were mostly German, but now and then some of ours were noticed also. The convoy arrived at B.82 at approximately 16:00 hrs.

The landing strip, which is situated just outside of Grave, consists of one grass runway without any metal tracking on it. The pilots found it very smooth but a little heavy due to recent rain. Last night the pilots who flew in with the aircraft slept in a barn. During this evening several of the personnel went around the country, in the nearby vicinity, and managed to arrange for sufficient rooms to put Squadron personnel under cover. The people seem very friendly and desirous of giving all the assistance possible.

October to December 1944

Operational flights soon began with patrols of the Nijmegen/Arnhem area on 2 October. Over the next couple of weeks, 402 gave this region special attention. Four days later during an afternoon fighter sweep, the Squadron encountered enemy aircraft and shot down three of them. The three successful pilots included future commanding officer F/L J.B. Lawrence. He recalls the action in his combat report:

402 SPITFIRE MK XIV OPERATING WITH 2 TACTICAL AIR FORCE

We were scrambled after Huns coming in from Venlo Wessel area. I was flying as Red 1 and led the squadron south.

On following the vectors from Kenway we came up underneath the Huns as they were approaching Nijmegen.

We climbed under a loose gaggle of 15 plus 109's when I sighted one 109 alone crossing in front of me. I turned into line astern and closed quickly. I fired one very short burst and the 109 went into a diving turn to starboard. I turned inside him and at about 20 degrees off 200 yds., range I fired another burst of about 2 secs. Strikes were observed on cockpit and engine. Pieces flew off, and white and black smoke poured out. The enemy a/c., turned into a steep spiral to port. He dived into the ground two or three miles S. of Nijmegen. I saw no parachute. After this engagement, five of us reformed and headed S.E. climbing. Red 2 and myself observed a lone 109 flying east at about 18,000 ft.

We head towards him and he went into a dive which steepened to the vertical. When near ground level, he attempted to level off. On pulling out, his aircraft disintegrated and fell in just N.E. of Cleve. I claim this for the squadron.

F/L A.R. BUD "REBEL" SPEARE IN THE COCKPIT OF HIS SPITFIRE MK XIV

F/L A.R. Speare was also a participant:

I was flying as Yellow 1 when Kenway reported Huns at 25,000 ft., S.E. of Nijmegen. Soon afterwards we sighted them at 12 o'clock above and we climbed up into them. I picked out two that were diving away. I lost one of them but followed the other Fw.190 from 20,000 ft. down to the deck.

The Hun took evasive action by doing steep climbing turns to starboard and then diving down again. I kept on his tail and took one 1 sec., burst when he was in a turn but did not allow enough deflection. I finally got to within 100 yds., range line astern and as he was doing a climbing turn I gave him a 2 sec., burst. I saw strikes on the engine, cockpit and wing and black smoke began to come from the e/a.

He half rolled to port and dove straight into the ground where he exploded.

The other pilot, F/O W.D. Whittaker, destroyed an Fw190. These significant victories ended a lengthy drought for 402. "This was the first occasion in many months in which the Squadron really had an opportunity to 'bag' the Hun and the boys lost no time in taking advantage of it." Even as they celebrated though, the fall weather began to interfere with the Squadron's operational flying.

402's aerodrome attracted the attention of the enemy. During an attack on 12 October, F/L K.S. Sleep, airborne at the time, thought one of the raiders he chased might have

KEN SLEEP READIES FOR FLIGHT. NOTE THE "SNUGGLE BUNNY" NOSE ART

been a jet fighter. During the pursuit, the enemy aircraft dropped two bombs on the side of the airfield where No.127 (RCAF) Wing was housed. No casualties were suffered but Sleep could not engage the mystery aircraft. Another attack followed the next day with the bandit positively identified as an enemy jet streaking over the airfield unopposed:

> Two bombs landed quite close to the Unit CO who was driving, with three other officers, just off the end of the field. Needless to say everyone hit the dirt. No casualties on 125 Wing but 127 again got the worst of it.[90]

F/O ARTHUR ERNEST "BARNY" BARNARD ON THE WING OF HIS SPITFIRE MK XIV

With the Luftwaffe showing up on concurrent days, sections of two aircraft were sent up during the morning of 14 October with hopes of catching them. Unfortunately nothing turned up for the pilots.

On 17 October, 402 mounted five patrols of the Nijmegen area. During one patrol the pilots saw white vertical trails. Later, the Canadians learned they were witnessing the launch of the new second-generation German revenge weapon, the V-2. These marked the last operational sorties for the Squadron before inclement weather settled in. Flying had to be postponed for a week, because heavy rain resulted in the airfield being declared unserviceable on 18 October.

B.82 again drew the attention of the Luftwaffe on 21 October, when enemy bombers attacked. No casualties were reported on the No.125 Wing side of the base but once more No.127 (RCAF) Wing bore the brunt of the bombing. Fifteen aircraft were damaged, with one airman killed and ten injured.

The weather cleared sufficiently on 24 October to allow the Squadron to carry out a successful armed reconnaissance. "The pilots were very pleased to fly after nearly a week on the ground and did good work on this operation damaging enemy rail communications."

The inclement weather affected even the ferry flights. F/O A.E. Barnard had been waiting for over two weeks at No.83 Ground Support Unit (G.S.U.) to fly a replacement Spitfire to Grave. No.127 (RCAF) Wing had to vacate the aerodrome on 22 October because backed-up water had flooded them out. No.125 followed suit three days later, being ordered back to Belgium on the 25th where they spent the winter at B.64 Diest. Heavy rain and a soggy field held up the Wing's departure for several days.

Finally, it cleared sufficiently on 1 November to allow both squadrons to transfer directly to B.64. The Winnipeg Bears were the last off the ground, and ran into deteriorating weather, which forced them to divert to B.78 at Eindhoven. This former Luftwaffe airstrip, although much better then what they left behind, did catch a few of the pilots off guard with its unfamiliar layout. "The strip here shook some of the pilots, due to its shortness and of the fact that they have to land uphill."[91] Even though the pilots became familiar with the new airfield, it still required caution during take-off and landing. "Due to the uphill nature of the strip and the hill at the end of it, landings and take-offs can only be made in one direction regardless of the wind."[92] The east-west runway often meant that pilots took-off directly into the sun, and the poor visibility caused by industrial haze only added to the challenge of B.78.

S/L J.B. Lawrence assumed command of the Winnipeg Bears on 29 October. A graduate of the British Commonwealth Air Training Plan (BCATP), Lawrence had marked time as an instructor in Canada before getting his wish for an overseas posting. His arrival at 402 was prefaced with stops at a Hurricane Operational Training Unit (O.T.U.) and 195 Squadron RAF, which operated Hawker Typhoons. His friends called him "Bud," a lifelong nickname bestowed by his father, and "Bambi," when they wanted to pull his leg. Lawrence rose through the ranks with 402, spending more than two and a half years with the Squadron.

Because of poor weather conditions, the Squadron only participated in five trips during November. The first operational flight from the new field at Eindhoven was an uneventful fighter sweep on 5 November. When they did see action, it involved attacking enemy ground communications and strafing motor transports. End of the month construction to improve the runway and taxi strips, rendered the airfield unserviceable.

The general summary for November, written by the Squadron's scribe, offers insight to how everyone felt: "It has been an extremely long and tedious month for the pilots, flying activity being almost at a standstill because of the weather and more so because of the poor conditions of the strip."[93] With the construction completed on 3 December, the airfield once again became serviceable. Operational flying began the next day and over the ensuing three days many patrols and a few uneventful armed reconnaissance flights were carried out.

41 Squadron joined the wing on 4 December, bringing the strength up to a full complement of three squadrons. Two armed reconnaissance sorties took place on 8 December, with the Squadron attacking a work party and the gates of an

unidentified canal. Weather continued to plague operations for the better part of the month, and the airstrip at Diest was declared unserviceable on 12 December due to the "extremely wet weather of the past few days."⁹⁴

The airfield construction crew spent the next four days laying down tracking and carrying out renovations. When the airfield maintenance was completed on 16 December, the Squadron took part in several armed reconnaissance sorties and sweeps. Poor weather continued to hamper operational flying with the frustrated pilots spending more time on the ground than in the air.

They woke up to a cool Christmas morning, but skies were clear and, despite a light tailwind, a successful armed reconnaissance was launched that afternoon. On the return journey, six aircraft landed at Brussels, where a taxi accident damaged two Spitfires. Flak forced one aircraft to land at Eindhoven. The rest of the formation returned to Diest where two separate incidents occurred on landing: one aircraft overshot the end of the runway, another crash-landed. "These crashes were in part due to poor conditions at Diest, the short runway is short and the visibility and wind conditions made landings very difficult."⁹⁵ As a result, five of the Squadron's aircraft were damaged beyond the capabilities of the servicing echelon, and were sent to No.409 Repair and Salvage Unit. The incidents and accidents of the day aside, a successful Christmas dinner was held at noon, with the officers serving the NCOs.

The officers dined later in the evening. F/O A.E. Barnard recalls the closeness of the Squadron that Christmas:

SQUADRON TRANSPORTATION PROVIDED BY THE CITY OF WINNIPEG; 402'S OWN 'CANADA' BUS

Late in the afternoon the Squadron was ordered to carry out a sweep and armed recce over the 'Battle of the Bulge' area. During the ensuing action Don Sherk got a Ju 88. (The 100th victory for 126 Wing since D-Day) and Bob Lawson was hit by flak. On the way back I was thinking; there goes our Christmas dinner, all we get is leftovers. We landed just at dusk and made our way to the mess. Surprise! Everyone had waited for us to return so that we could all eat together – a true 'Band of Brothers'.

In 2005, 402 initiated the *return of the late arrivals* to commemorate that Christmas dinner in 1944. This became a new Squadron mess dinner tradition. It should be noted that

F/L (Ret'd) A.E. Barnard was also in attendance the at the inaugural mess dinner, 62 years later.

402 SQUADRON PILOTS, DIGBY, LINCS, JANUARY 1944 – STANDING L TO R: ART "BARNY" BARNARD, FRED D. MILES, HOWARD C. "NICK" NICHOLSON, WILLIAM M. "BILL" BURNETT, NICHOLAS B. "DIXIE" DIXON, WALTER N. "DICK" SHEPHERD, UNKNOWN – KNEELING: JAMES M. MCMAHON, KENNETH S. SLEEP, MARTIN A. NAYLOR, UNKNOWN

Art Barnard Collection

The Squadron moved on to B.88 Heesch in Holland to join their fellow Canadians of No.126 (RCAF) Wing. 402 welcomed this news. "Aside from a slight amount of confusion and delay in settling down the Squadron is now quite 'at home' on 126 Wing and is particularly happy to at last be part of a Canadian Wing."[96] The advance party, along with the unit's aircraft, left on the 27th. Fog, sleet and mechanical problems with some of the vehicles prevented the groundcrew from leaving until the 29th.

The Winnipeg Bears flew several uneventful operational flights by month's end.

January to July 1945

On New Year's Day 1945, about thirty Fw190s from Jagdgeschwader 6 attacked the aerodrome at Heesch at 0910 hours.[97] This group was part of a larger force of some 900 German aircraft, taking part in Operation Bodenplatte. Hitler was playing his last desperate card in an attempt to both cripple allied air forces in the Low Countries, and keep the Wehrmacht on the attack during the Ardennes Offensive. 465 Allied aircraft were damaged or destroyed by the Luftwaffe during Bodenplatte.

At Heesch, the German aircraft swept over the strip, firing a few bursts before continuing on to Eindhoven, where they caused considerable damage. Unfortunately, a shortage of aircraft that morning made it impossible for 402 to get off the ground and respond. By the afternoon, they were able to launch eight uneventful patrols of two aircraft each over Nijmegen and Volkel, before weather cancelled flying for the next two days.

When the skies cleared on the afternoon of 4 January, the Squadron sent pairings of Spitfires up to patrol. 402 followed this up with an armed reconnaissance flight, which attacked rail transport "and the pilots thought it a good show." The Winnipeg Bears were beginning to enjoy their new home, a big improvement over the tricky Diest.[98] "It is made of the American style tracking, and has the tracking on all taxiing [sic] strips as well as the bays, and is a tremendous improvement over the conditions we operated under at Diest."[99] Poor weather meant very little operational flying over the next ten days and little enemy contact.

Eighteen aircraft departed for RAF Station Warmwell, Dorset, to attend No.17 A.P.C. (Armament Practice Camp) on the 15th. The pilots spent the rest of the month training on an air-firing and dive-bombing course. Fifty of the groundcrew followed several days later. All told, the Squadron put up 150 flights at the camp. Their time at Warmwell was almost incident free with only one aircraft being slightly damaged. A return flight to Heesch on 1 February brought about several prangs due to deteriorating weather and poor visibility. The pilots, unable to land at Heesch, were diverted to B.80, Volkel, Holland, where one aircraft crashed on landing. Engine failure caused another Spitfire to crash near Bourg Leopold in Belgium.

The armament training at Warmwell served the Squadron well. The noose around the retreating Wehrmacht was getting tighter, and over the next three months, numerous ground targets of opportunity were available. The Luftwaffe lost many irreplaceable pilots and never recovered from their disastrous operation on 1 January. Although sightings and engagements of German aircraft were becoming a rarity, there continued to be the occasional contact. For example, on 8 February, F/L Ken Sleep registered his first victory. This successful sortie also marked the happy return of newly promoted W/C G.W. Northcott, DSO, DFC and Bar, who once again led 402 as part of 126 Wing.

The Squadron engaged in armed reconnaissance sorties over the next few weeks and saw some action. Now equipped with bombs, 402's Spitfires had the opportunity to make a more significant contribution during their armed reconnaissance operations. Their first bombing sortie took place on 22 February, the same day S/L L.A. Moore, DFC, AFC, took over command of the Squadron. Two more shows followed on the 25th. Remarkably, during the first op, F/L's K.S. Sleep and B.E. Innes shared in damaging an Me262, the world's first jet fighter. Sleep's combat report relates the engagement.

S/L LES MOORE

I was leading Black section of 402 Squadron on a dive bombing mission to the Bocholt area. After bombing we continued on an armed recce to the Twente area.

We had circled the TWENTE A/D and had turned on a course of about 270 degrees at 6,000 feet when a Me.262 with light green camouflage flew through us, head on. We turned to engage but were unable to close and it disappeared heading towards Lingen.

We resumed our course to base when sighted another Me.262 coming head-on. We broke into the Me.262 and I fired a 2-3 second burst from 400 yards, 30 degrees starboard and behind, as he was turning to port. A few pieces fell off the e/a but I was unable to tell exactly from what section. F/L Innes also had strikes on this a/c, but we were unable to close sufficiently for further combat.

On the second show, F/L W.S. Harvey's aircraft was hit by flak from the Enschede area, forcing him to bail out. Harvey became a PoW but later escaped.

Maintaining any type of combat aircraft in the field is difficult work. Because of the number of sorties and the type of flying performed with the Squadron's Spitfire XIVs, the ever-present gremlins took their toll on both the groundcrews and servicing echelon. However their dedicated work paid off on 4 March with the acknowledgment that 402's groundcrew had the highest serviceability of all the Spitfire XIV Squadrons in No.83 Group.

A FITTER HARD AT WORK ON A MERLIN

A memo dated 5 March forbid further attacks on ground targets. Thus, for the next few weeks, the Squadron engaged in sweeps seeking out enemy aircraft.

The new Me262 jet was a highly sought after but difficult and elusive target. On 13 March the unit took part in two shows, the first involved providing an escort for medium bombers over the Rhine area. In the afternoon, a Sweep of the Munster, Bielfeld, Hamm and Dorsten area resulted in F/O H.C. Nicholson downing an Me262 from I./KG 51, that was engaged in attacking the bridge at Remagen.[100]

> I was flying Yellow 3 on a fighter sweep in the Gladbach area when I sighted a Me.262 at about 5000 feet flying South West. He did not appear to see me. I broke and fired a 3 second burst from 250 yards line astern into his starboard wing. I kept on firing, observing many hits and the aircraft tended to fall out of control, regaining slowly. At 2000 ft., he went into a sharp dive to port but owing to the extremely heavy flak from Gladbach, I broke to starboard. I did not see him crash, but this is confirmed by the C.O. of 402 Squadron.

S/L Moore described the action:

> When F/O Nicholson attacked jet a/c, I saw strikes and pieces fly off a/c and when F/O Nicholson broke off his attack, I followed e/a down to 500 ft., and fired, but saw no results. E/A was out of control at that time in a vertical dive.
>
> I saw a large flash on the ground but was unable to pay much attention because of intense flak.

There were several more bombing trips, escort missions and armed reconnaissance operations. These involved very little action for the pilots until 25 March when the Commanding Officer, S/L L.A. Moore was lost during an armed reconnaissance.

THE SQUADRON WHILE COMMANDED BY S/L LESLIE MOORE, (CENTRE ROW, FIFTH FROM LEFT)

"It was extremely unfortunate that during an attack on a locomotive on March 25th our O.C. S/L L.A. Moore was seen to dive straight into the deck, apparently hit by flak or flying debris. His loss is sincerely regretted as S/L Moore was respected by everyone and was unquestionably a good leader."[101] Moore was well known in the Squadron through a previous posting from January 1943 to March 1944. A six-victory ace, his bravery had been recognized through the awards of the DFC and the AFC. Born in Hamilton, Ontario, but raised in Plainsfield New Jersey, U.S.A., the 23-year-old, rests in Reichswald Forest War Cemetery, Kleve, Germany.[102]

The Winnipeg Bears encountered more enemy aircraft during the last days of March. On the 30th, the unit carried out four armed reconnaissance flights. During the third show, F/L H. Cowan shot down an Fw190. The next day brought the same number of armed reconnaissance sorties over the front line. During the first op, F/O R.W. Lawson quickly disposed of two Fw190s.

402 SPITFIRE MK XIVs

I was flying Yellow 3 in 402 Squadron on Sweep/Armed Recce when I saw two aircraft passing under our section. Heading S.W. I called Red 1 and half-rolled down after the aircraft. I recognized them to be Fw.190's [sic]. I picked out No.2 and fired from 150 yds., a second and half burst from approximately 10 degrees port and slightly above.

I saw strikes and a burst of flame from near the cockpit. He then half-rolled into a woods and exploded. I closed to 200 yds., on the leading enemy a/c and gave a half second burst, saw no results. Then moved to 100/150 yds., deadline astern and fired one and half second burst, saw strikes and the jet tank and coop-top fly off, a burst of flame, the enemy a/c then did a slight wing over and exploded in a field.[103]

A few minutes after Lawson's combats, F/L B.E. Innes also destroyed an Fw190.

Meanwhile the Squadron continued their busy pace during April. A strong crosswind cancelled most flying for the first couple of days, but one uneventful patrol took place over the Enschede, Borken, Coesfeld area on 3 April. Two days later, continuous four-man patrols were maintained along the Rhine. During the first patrol, F/L E.R. Burrows was leading a six-plane formation when they ran into a mixed gaggle of approximately 20 Fw190s and Me109s. The pilots performed admirably with a score of two destroyed, four damaged and one probable. F/L H. Cowan was involved in this melee, damaging an Fw190 and probably destroying another. He describes the action:

When on patrol over Lingen on course 350 degrees we sighted about 20 Fw.190 and 109 e/a. We broke into them to starboard, they broke port and we followed. I chased one Fw.190 into cloud and subsequently lost him. I and my No.2 came below cloud and circled. A Fw.190 broke cloud behind us during our second turn and I pulled up after him decreasing our range until the e/a fitted well into my Gyro Sight (aircraft gun sight). On opening fire from line astern and long range (about 700 yards) I got in a 2 to 4 second burst seeing strikes all over the cockpit and wingroot, whereupon he started pouring thick black smoke and flicked to starboard and vanished into a cumulus cloud in a downward turn. I looked around to see if all was clear to follow him when I sighted a second 190 sitting right over the top of my a/c. I pulled up steeply and was closing too fast and steeply to use Gyro Sight. I just lined e/a up in the glass of the sight and fired from about line astern and 50 yards, observing strikes on both wings. E/A spun to starboard and I pulled my a/c to port to avoid him as my speed had dropped to less than 100 mph.[104]

Patrols of the Lingen, Rhine area were maintained the next day, and although no enemy aircraft were encountered, F/L H.L. Murray was wounded by flak during his return to base. After safely landing his Spitfire, Murray lost consciousness. S/L Moore's replacement arrived that same day. S/L D.C. Laubman, DFC and Bar, assumed command on 6 April. Flak, one of the most dangerous aspects of attacking ground targets, had begun to take its toll. During a patrol on 11 April, when the Squadron claimed 10 M.E.T. (Motorized Enemy Transport) 1 Bus and 2 trailers, P/O G.F. Peterson was lost to flak. His aircraft crashed near the Deelen Airfield north of Arnhem. George Frederick Peterson, a 22-year-old from Toronto, is buried in the Oosterbeek War Cemetery, Arnhem, Holland.[105]

The following day, 402 transferred to a temporary base, B.108, situated on the Rhine. Although only staying for a brief three days, the Squadron lost another commanding officer. On 14 April, S/L Laubman led twelve aircraft on an armed recce into the Salzwedel, Soltau, Zeven areas.

During an attack on a pair of enemy vehicles, Laubman's rounds found their mark on the rear one which turned out to be a fuel vender. His Spitfire flew through the ensuing fireball and began losing glycol before the engine seized.[106] Laubman bailed out and was rescued by the German soldiers after a confrontation with angry German civilians and members of the Hitler Youth. He spent the remainder of the war as a PoW.[107] His tour as 402's O.C. lasted only nine days, but Laubman, one of Canada's great aces of the war with 14 victories,[108] went on to have a lengthy and successful postwar Air Force career, retiring with the rank of Lieutenant-General in 1972.

A day after losing Laubman, the Winnipeg Bears were on the road again to their new base, B.116, a former Luftwaffe airfield in Wunstorf, Germany. The transfer of the Spitfires turned out to be more than a simple ferry flight. Two armed recces of six aircraft each were mounted, resulting in damage to some rail and road transportation en route to their new home.

ONE OF THE BEST KNOWN COMMANDERS OF THE SQUADRON, S/L DON LAUBMAN

402 Squadron Archives

THE GUARDHOUSE AT B.116 WUNSTORF AERODROME

402's new skipper, S/L D.C. Gordon, DFC and Bar, hailed from Edmonton. Donald Campbell "Chunky" Gordon's dossier included desert fighting on Hurricanes with 274 Squadron, many hours on Spitfires with 417, 403, 442 and 411 squadrons, and eleven victories to his credit.[109]

S/L "CHUNKY" GORDON INSPECTS WRECKED TIGER TANK IN THE RUINS OF AACHEN, GERMANY, 1945

There was no rest for the Winnipeg Bears during this hectic period. On 16 April, three armed recces and two patrols, amounting to thirty-five sorties, were launched. From these actions 402 chalked up thirteen M.E.T., (mechanized enemy transport) fifteen locomotives, damaging twenty-six T.R.G.S. (railway cars) and one building destroyed. Although F/L J.E. Maurice failed to return from one of these missions, he survived, became a PoW and eventually made his way back to the Squadron.

402 became involved in a different kind of operation on 17 April. Instead of destroying bridges, they maintained their safety by continually carrying out patrols of four aircraft over bridges along the stretch of Celle to Nienburg.[110]

The first mission, led by the new O.C., S/L D.C. Gordon, took place on 18 April. His formation caused considerable damage to locomotives, T.R.G.S., M.E.T. as well as buildings in the Goldberg/Wittstock area. Flak claimed F/L H. Cowan while chasing an enemy aircraft over the Luftwaffe Parchim aerodrome. The following was noted in the ORB: "He was not seen to bale [sic] out and burning wreckage was visible on the drome." It also contained a brief epitaph about Cowan – "Always intrepid and very keen, his loss is sincerely regretted." 24-year-old Henry "Hank" Cowan, a Jew who had escaped from Germany in 1938 and settled in Trout River, B.C., is buried in the Berlin War Cemetery, Charlottenburg, Germany.[111]

402 returned to action on the 20th, carrying out three armed recces in the Wismar, Lubeck, Kiel, Meldorf and Hamburg areas. During the first show, Warrant Officer (WO) V.E. Barber attacked a locomotive but had to bail out after his Spitfire suffered damage from debris from the target. During the second sortie that day, S/L Gordon led a team of seven aircraft. 401 Squadron invited the Winnipeg Bears to join a free-for-all resulting in the destruction of a pair of Fw190s. F/O T.B. Lee, who claimed one of the Focke-Wulfs, describes the action:

> Flying in a direction N.W. of Hagenow Aerodrome at about 500 feet, F/O (A.G.) Ratcliff, 402, Yellow one, sighted it first and went in on the attack. He fired several bursts in a running fight and broke away. I, 402 Yellow 2, pressed into attack from 400 yds., and after several steep turns near the ground pressing the attack to 200 yds., I obtained hits in the cockpit which produced a sizeable flame. The aircraft proceeded into a German house and exploded, spreading fire all over the immediate vicinity.[112]

Five front line patrols took place on 21 April and included another encounter with enemy aircraft. During the engagement, F/L E.R. Burrows (known around the Squadron as "Rabbit") shot down an Me109. The German pilot bailed out but his parachute failed to open. F/O H.G. Dutton and F/L W.O. Young also damaged two more Me109s.

SPITFIRE RN119 AE J, ON A PSP-COVERED DISPERSAL, MARCH 1945, B.88 HEESCH

Low cloud and rain showers cancelled flying over the next couple of days, but the Squadron returned to patrolling on the 24th. Two armed recces of the Hitzacker, Wismar, Kiel, Hamburg area were launched on 26 April, and the second show, led by F/L Burrows, yielded success. They attacked aircraft moored on water at the Putnitz aerodrome: F/L B.E. Innes destroyed one He115 and F/O R.H. Roberts damaged another. Ground targets were also strafed during the second show including a pair of locomotives and some M.E.T. damaged. A return engagement at the Putnitze aerodrome the next day netted similar results. Two He115s were destroyed, one damaged, plus two locomotives and one M.E.T. were shot up and claimed as damaged.

Despite the winding down of the war, the Allies kept unrelenting pressure on the retreating Germans. 402 continued to carry out patrols in hopes of engaging the remnants of the Luftwaffe, but for the time being had to be content with gathering vital information from armed recces.

The Winnipeg Bear's luck changed dramatically at the end of the month. The last major engagement of the war for the Squadron on 30 April turned out to be the grande finale. During the first show, a section led by F/L D.R. Drummond, ran into a concentration of German aircraft. The net result showed eight enemy aircraft destroyed, four damaged, and one locomotive damaged. During the second show of the day, F/L E.R. Burrows' section shot down one Ju88, and damaged two others, in addition to six enemy vehicles destroyed. A third patrol, also lead by Burrows, proved uneventful, but it had been quite a day for the Canadians.

On 1 and 2 May, 402 took to the sky again for some uneventful patrols in the area from Ahrensburg to Zareentin. The next day, three armed recces were carried out over the Bad Segeburg, Kiel, Rendsberg, Elmshorn and Hamburg areas. During the second show, three Fiesler Storch aircraft were thoroughly strafed on the ground. Shortly after, S/L Gordon caught one flying low off the deck and made short work of it as detailed in his combat report:

> I was leading 402 Squadron on an Armed Recce of the Kiel area, when at 4,000 feet West of Neumunster my No.2 reported an a/c flying near the deck. He went down and made an attack on the a/c, which he identified as a Fiesler Storch, and missed it completely. I followed him in and opened fire at 400 yds, at 90 degrees deflection seeing strikes all over the e/a. The aircraft caught fire in the air and crashed in flames.

Ground targets also came under fire during this recce. F/L J.A. O'Brian's Spitfire, disabled from target debris, left him no choice but to bail out. By the time the Squadron returned from this sortie, O'Brian had already contacted base to say that he was on his way back and he returned the next day. O'Brian has the unfortunate distinction of being the last member of the Squadron to bail out of an aircraft during an operational sortie.

Twelve aircraft, on a sweep into southern Denmark, attacked and destroyed two M.E.T. and damaged ten more on 4 May. During this operation, an Me262 attacked F/O J.E. Rigby and shot up his Spitfire. Armed with four 30mm cannons, the German jet fighter fired devastating exploding ammunition, and Rigby was fortunate to have escaped uninjured. This marked the final day the Squadron took part in an operational activity, and the cessation of hostilities on the front at 0800 hours on 5 May, cancelled all flying. Later that day, three members of the Squadron who were lost on operations returned to the fold. S/L D.C. Laubman, F/L J.E. Maurice and WO1 V.E. Barber were all happy to be home.

The end of hostilities and unconditional surrender of Germany on 8 May turned the next two days into a Victory in Europe celebration for both the Squadron and Wing. Other than occasional practice flights to keep the pilots and groundcrew in check, little flying took place. The Squadron received word that they were to move to B.152 Fassberg on 11 May. They completed the move three days later and stayed at B.152 until the end of June. Events had wound down to such a point that the disbandment of some of the Squadrons (including 402) and Wings could begin.

It was learned towards the end of the month that No.126 Wing was to be the Wing selected for the Occupational Forces. No.401 and 402 Squadrons were transferred to No.127 Wing and all Canadian personnel in No.83 Group with the highest priorities for repatriation or who wished to be in the occupational Forces or serve in the Pacific Theatre of war would be posted to disbanding units.

On 25 June, the Squadron exchanged their Spitfire XIV's with 412 Squadron's Spitfire XVIs. "This unit traded Mk.XIV Spitfires with 412 Squadron, receiving in place of them Mk.XVI Spitfire aircraft. No.412 Squadron is one of the units that will be making up the Canadian Occupational Wing in Germany."[113] The Spitfire Mk XVI was essentially a Mk IX with an American built Packard Merlin engine. After flying the powerful Mk XIV, the pilots agreed that the difference in power and climbing ability was clearly evident.

On 2 July, the Squadron transferred to No.127 Wing where they remained until disbandment on 10 July 1945. The disbandment, which lasted less than a year, marked the only stand down during the long history of 402 "City of Winnipeg" Squadron.

The Squadron's proud war record includes: 10,504 sorties. Operational/Non-operational flying hours: 17,643/12,027. Victories: aircraft 49 destroyed, 10 probably destroyed and 37 damaged, plus 5 V-1 (flying bombs) destroyed. Casualties: Operational 47 pilots of whom 36 were killed or missing, 3 POW, 1 killed, 1 evaded capture, 7 wounded. Non-operational: 12 personnel killed, 1 injured. Honours and Awards 1 Bar to DFC, 4 DFCs. Battle Honours: Defence of Britain 1941-1944. English Channel and North Sea 1941-1944. Fortress Europe 1941-1944: Dieppe, France and Germany 1944-1945: *Normandy 1944, Arnhem, Rhine.*[114]

Hurricane 1940

Just twisted scrap thrown on a dump
Strips of wing and a Merlin sump
Old fighter plane
Your flight is done
Your landings made and victories won

Gun barrels scorched and motors tired
Your masters fought as men inspired
Old fighter plane
They trusted you
Who faithfully served the Gallant Few

Casually now they fly around
Jet propelled at the speed of sound
New fighter planes
Fierce in your power
Spare thought for those
Who had their hour

The Right Honourable Lord Balfour of Inchrye, P.C, M.C.,
Under Secretary of State for Air Great Britain

NOTES

1. 181.002(D266).
2. Darman, Peter. *World War II Day by Day*, p. 22.
3. 402 Operations Record Book (ORB).
4. 402 ORB.
5. 402 ORB.
6. Darman, Peter. *World War II Day by Day*, pp. 48-49.
7. 181.002(D266).
8. Ramsey, Winston G. *Battle of Britain, Then and Now*, p. 253.
9. 402 ORB.
10. 402 ORB.
11. 402 ORB.
12. 402 ORB.
13. 402 ORB.
14. 402 ORB.
15. 402 ORB.
16. 181.022 (D1335) Formation of Further RCAF Squadrons, March 1, 1941.
17. 402 ORB.
18. "Sweep" – Offensive formation of fighters or fighter bombers over enemy territory, designed to draw the enemy. See Franks, N.L.R. *Royal Air Force Fighter Command Losses of the Second World War*, Volume 2, p. 9.
19. Titles or nicknames "City of Winnipeg Squadron" and "Winnipeg Bears" – S. Kostenuk and J. Griffin. *RCAF Aircraft and Squadrons*, p. 84.
20. A.D.G.B.
21. Allison, Les and Harry Hayward. *They Shall Grow Not Old*, p. 578.
22. Bishop, Arthur. *Courage in The Air – Canada's Military Heritage Volume 1*, p. 134.
23. Franks, N.L.R. "Rodeo – Fighter sweep without bombers", *Royal Air Force Fighter Command Losses of the Second World War*, Volume 2, p. 9.
24. Kostenuk, S. and J. Griffin. *RCAF Aircraft and Squadrons*, p. 85.
25. Franks, N.L.R. "Cat Ac, damaged, repairable on site but not by the operating unit - Cat B, damaged, repairable at a maintenance unit, civilian repair depot, or manufacturer", *Royal Air Force Fighter Command Losses of the Second World War*, Volume 2, p. 9.
26. A.D.G.B., p. 78.
27. Bowyer, Chaz. "The Hurribomber could also be loaded with two 500lb bombs; a configuration used during the Dieppe raid", *Hurricane At War*, p. 73.
28. Franks, N.L.R. "Circus - Bombers heavily escorted by fighters to bring enemy fighters into combat", *Royal Air Force Fighter Command Losses of the Second World War*, Volume 2, pp. 7 & 9.
29. Allison, Les and Harry Hayward. *They Shall Grow Not Old*, p. 428.
30. "Operation Channel Stop" - the code name given to operations designed to prevent German surface units from passing through the English Channel during daylight, <www.rafweb.org>.
31. "Ramrod" – Similar to a Circus, but with intention of destroying a target. See Franks, N.L.R. *Royal Air Force Fighter Command Losses of the Second World War*, Volume 2, p. 9.
32. 402 ORB.
33. 402 ORB.
34. Combat Report.
35. Allison, Les and Harry Hayward. *They Shall Grow Not Old*, p. 571.
36. Darman, Peter. *World War II Day by Day*, pp. 120 & 121.
37. *The Marshall Cavendish Illustrated Encyclopedia of World War II – Volume 6*, p. 836.
38. *The Marshall Cavendish Illustrated Encyclopedia of World War II – Volume 6*, p. 835.
39. Combat Report - Form F.
40. Allison and Hayward, p. 205.
41. Ibid., p. 212.
42. Air Defence of Great Britian (A.D.G.B.), p.103.
43. Ibid.
44. "Super Circus". See Franks, N.L.R. *Royal Air Force Fighter Command Losses of the Second World War*, Volume 2, p. 9.

45 A.D.G.B., pp. 105-106.
46 402 ORB.
47 A.D.G.B.
48 "Rhubarb" – Small-scale freelance fighter sorties against ground targets of opportunity. See Franks, N.L.R. *Royal Air Force Fighter Command Losses of the Second World War*, Volume 2, p. 9.
49 A.D.G.B.
50 Price, Dr. Alfred. *Late Marque Spitfire Aces 1942-45*, p. 9.
51 Bishop, Arthur. *Courage in the Air, Canada's Military Heritage, Volume 1*, p. 115.
52 402 ORB.
53 *JG-26 War Diary - Volume I 1939-1942*, p. 281.
54 Allison and Hayward, pp. 448 and 523.
55 <www.bbc.co.uk/ww2peopleswar/stories>.
56 A.D.G.B., p.130.
57 Ibid.
58 Allison and Hayward, pp. 223 and 739.
59 "Roadstead" – Low-level attack on coastal shipping. See Franks, N.L.R. *Royal Air Force Fighter Command Losses of the Second World War*, Volume 2, p. 9.
60 402 ORB and Form F.
61 Allison and Hayward, p. 113.
62 "Jim Crow" – A fighter recce sortie over English Channel. See Franks, N.L.R. *Royal Air Force Fighter Command Losses of the Second World War*, Volume 2, p. 9.
63 <www.german-navy.de>.
64 Salmaggi & Pallavisini. *2194 Days of War.*
65 Allison and Hayward, p. 16.
66 Ibid., p. 537.
67 *RAF in Maritime*, p. 501.
68 "Lagoon" – Anti-shipping operations in company with Coastal Command Beaufighters. See Franks, N.L.R. *Royal Air Force Fighter Command Losses of the Second World War*, Volume 2, p. 9.
69 402 ORB.
70 A.D.G.B.
71 British Intelligence, p. 116.
72 402 ORB.
73 Squadron war diary
74 402 ORB.
75 <www.learningcurve.gov.uk>.
76 402 ORB.
77 402 ORB.
78 402 ORB.
79 Allison and Hayward, p. 134.
80 Ibid., p. 569.
81 402 ORB.
82 Rawlings, John D.R. *Fighter Squadrons of the R.A.F. and their Aircraft*, p. 423.
83 402 ORB.
84 British Intelligence in the Second World War, p. 534.
85 402 ORB.
86 Consolidated Diver Report No.402/S Hawkinge, August 16, 1944.
87 Consolidated Diver Report No.402/S/4, August 23, 1944.
88 A.D.G.B., pp. 325 & 326, and 402 ORB.
89 402 ORB.
90 402 ORB.
91 402 ORB.
92 402 ORB.
93 402 ORB.
94 402 ORB.
95 402 ORB.
96 402 ORB.
97 *Bodenplatte - The Luftwaffe's Last Hope*, p. 125.
98 402 ORB.
99 402 ORB.
100 *Me.262, volume three*, p. 603.
101 402 ORB.
102 Allison and Hayward, p. 534.

CHAPTER 2

103 combat report.
104 combat report.
105 Ibid., p. 599.
106 402 ORB.
107 Bishop, Arthur. *Courage in The Air*, p. 198.
108 Spick, Mike. *Allied Fighter Aces*, p. 216.
109 Shores and Williams. *Aces High*, pp. 292 & 293,.
110 402 ORB.
111 Allison and Hayward, p. 147.
112 combat report.
113 R.C.A.F. O/S H.Q., London, Information Memo No.133.
114 Kostenuk, S. and J. Griffin. *RCAF Aircraft and Squadrons*, p. 85.

402 Squadron War Record

SORTIES: 10,504

OPERATIONAL/NON-OPERATIONAL FLYING HOURS: 17,643/12,027

VICTORIES:

    Aircraft: 49 Destroyed

                  10 Probably Destroyed

                  37 Damaged

    5 V-1 (Flying Bombs) Destroyed (All in August 1944)

CASUALTIES:

    Operational: 47 pilots, of whom 36 were killed or missing

    PoW: 3 (1 Evaded capture)

    Wounded: 7

    Non-operational: 12 killed, 1 injured

HONOURS AND AWARDS:

    1 Bar to DFC, 4 DFC's

BATTLE HONOURS:

    Defence of Britain 1941-1944

    English Channel and North Sea 1941-1944

    Fortress Europe 1941-1944: *Dieppe*.

    France and Germany 1944-1945: *Normandy 1944, Arnhem, Rhine*.

Source: *RCAF Squadron and Aircraft* - S. Kostenuluk and J. Griffin

A LOVELY COLOUR STUDY OF THE SQUADRON'S MUSTANGS ON THE RAMP AT RCAF STATION WINNIPEG

John Meush Collection

402 Squadron, 1946-1968

Dr. Leo Pettipas

Introduction

With the cessation of hostilities, all of the RCAF bomber, coastal, and fighter squadrons were disbanded pending a decision as to the type of peacetime air force Canada should have. In the spring of 1946, the RCAF was busy preparing three plans for the government on the structure of the postwar RCAF. After six years of war, the federal coffers were seriously depleted, and what the government was looking for was a credible defence force for as little money as possible. However, although the war had been over for less than a year, it was clear that the country would have to retain a military establishment into the postwar era. Various Allied conferences and battlefield experiences during the war had shown that serious incompatibilities existed between the Soviet Union and the West. Prime Minister Mackenzie King anticipated that the USSR and the U.S.A. would someday be in contention, placing Canada squarely in the middle. In particular, Canada's northern real estate was perceived by the Americans as a potentially easy line of Soviet advance into the "Lower Forty-Eight." They regarded all northern roads and airfields as defence assets of continuing strategic importance, and it was feared that, if Canada did not undertake defence measures on a scale the Americans deemed adequate, they might infringe on Canadian sovereignty to whatever extent they felt necessary to ensure continental security.

On 22 February 1946, the Minister of National Defence announced a postwar plan for the RCAF that provided for a Regular force of 16,100 officers and airmen, an Auxiliary force of 4,500, and a Reserve of 10,000. It was proposed to have eight squadrons in the Regular force and fifteen in the Auxiliary. This harkened back to the pre-war RCAF in which the Auxiliary had played an active role, not only in providing personnel, but also in mobilizing a military force for the war.

The No. 2 Air Command Years (April 1946 - March 1947)

On 15 April 1946, 402 Squadron was re-activated at the site of the former No.5 Air Observer School, Stevenson Field, under the control of No.2 Air Command, a temporary organization formed in late 1944 to cover the closing down of the British Commonwealth Air Training Plan (BCATP) and the establishment of the peacetime command structure. 402 Squadron was one of the Auxiliary units that would comprise a first-line reserve of fully organized, manned and equipped squadrons that could be mobilized on short notice. Originally, the Squadron was to be a fighter-reconnaissance unit equipped with de Havilland Mosquitoes. On 1 May 1946, the organization order was amended to specify a fighter-bomber function under the designation "No.402 (FB) Sqn (Aux)." This proved to be the first of several administrative and role changes that the Squadron would experience during its ensuing career. In the latter half of the 1940s, the perceived Soviet threat was twofold – land invasion, and intrusion into North American airspace by heavy bombers of the Soviet Long Range Air Force. Changing views as to the relative seriousness of these two threats produced periodic role changes for 402 Squadron during the first decade of its existence. In March 1947, its mandate was converted to that of a straight fighter unit and, on 1 April 1949 it was changed again, this time back to that of fighter-bomber.[1]

402's first postwar CO was W/C R.J. Clement, DFC, an employee of the Engineering Department of Trans-Canada Air Lines and a pilot in civilian life who took command on 2 August 1946. This appointment was a homecoming for W/C Clement, who had previously served with the Squadron in 1937.

W/C R.J. CLEMENT, DFC, CD

THE DE HAVILLAND MOSQUITO – 402'S RUMOURED AIRCRAFT IN 1946

A key element of 402 Squadron was its Regular Support Unit (9402 RCAF "R" Detachment) that was mandated "to advise and assist the CO of 402 (FB)[2] Squadron (Auxiliary), Winnipeg, in administration, maintenance, and air and ground training of personnel."[3] Indeed, the first few months of the postwar history of 402 were more a story of 9402 than anything else. The first members of the Detachment reported in on 1 May 1946, and work was taken in hand to clean up Nos. 2 and 3 Hangars, procure synthetic training equipment from sites as far away as Lethbridge and Calgary, landscape the grounds, and paint building walls. The first aircraft, Harvard AJ 832, arrived on 23 May 1946, and the build-up toward the Squadron's Unit Establishment (UE) of nine such machines proceeded over the following months. The Harvards were acquired for refresher and operational flying training in anticipation of eventually receiving combat-type aircraft as well. The Auxiliary syllabus allowed for 200 hours of ground instruction and 125 hours of flying instruction, with the first "business" flight occurring on 4 June 1946. Once the Auxiliary program was in place, "parading" would take place two nights a week and on weekends.

An event that would be repeated annually over the ensuing years was the staging of Air Force Day. It first took place on 12 July 1946, and among the aircraft featured were a pair of Hawker Hurricanes and two Curtiss Kittyhawks that had survived the late war and had not yet been sold off by the government. Of a rather more serious nature was a mercy mission flown by two Detachment personnel: on 24 August, 402 was tasked to fly a badly injured man from Melfort to Regina, Saskatchewan. The Regina-based civilian ambulance aircraft was unserviceable, so F/L P. Bissky, pilot, and FSgt E. Vandahl, co-pilot, took off from Winnipeg in an Avro Anson communications aircraft in the wee hours of the following morning. After refuelling at Yorkton, the aircraft carried on in difficult weather conditions to Melfort where the man, suffering from a fractured neck, was placed aboard the Anson. He was conveyed to the Regina airport and transported to a city hospital by ambulance. This mercy mission was the harbinger of many others the Squadron would undertake decades later.

In September, the RCAF was flying lend-lease aircraft from storage sites in Saskatchewan back to the United States. The pilots would return to Canada in a No.124 (Ferry) Squadron Douglas Dakota transport, and on one such return trip, the aircraft crashed into a gully on final approach to the Estevan airport, killing all on board. This was the second-worst single-aircraft accident in the history of the RCAF.[4] Later that month, it fell to pilots of 9402 Detachment to fly the remains of several of the deceased to locations in Western Canada for burial. Back at the station, the Air Officer Commanding

(AOC) visited the unit and volunteered the information that the Squadron's Mosquitoes would soon be arriving. He directed that the assemblage of display aircraft housed in No.2 Hangar, destined for use as the maintenance facility, be removed forthwith so that refurbishment of the building could begin. The aircraft – Ansons, Bolingbrokes, Hurricanes, Kittyhawks, and a Norseman – were made serviceable and flown to the aerodrome at Portage la Prairie for storage. An "instructional" Mosquito finally arrived in mid-October, but in the end it was decided that this type of aircraft would not form the Squadron's operational complement.

Meanwhile, interviews were carried out with prospective Auxiliary air and ground personnel, and by mid-November, refresher training using the Squadron Harvards was underway. Between December 1946 and May 1947, exploratory flights were made over the Interlake and the countryside west of Lake Manitoba in search of suitable low-flying areas, and the Gimli aerodrome was selected as the site for forced and precautionary landing practice.

A 402 HARVARD UNDER POWER AT GIMLI – NOTE THE LINE-UP OF VAMPIRES AND HARVARDS

Al Chalkley Collection

On 11 December 1946, the late-Pilot Officer Andrew Mynarski was posthumously awarded the Victoria Cross for his efforts to save a fellow crewman's life in a burning Lancaster bomber during the war. F/L R.H. Dibnah of 9402 Detachment was an attending officer at the investiture ceremony in which the medal was presented to P/O Mynarski's mother. In a year marked by "firsts," Auxiliary personnel commenced their initial minor inspection on a Harvard aircraft on 17 December.

1947

The new year brought with it more flying activities appropriate to a Squadron that was gathering momentum: familiarization flights, local circuits, solo and night flying checks, stalls and spins, aerobatics, instrument flights, clear-hood dual, formation flying, simulated forced-landing practice, progress checks, night and cross-country navigation exes, non-radar tower-controlled interceptions, and flare-dropping exercises over Lake Winnipeg. The latter were carried out to prepare the pilots for the use of such devices to illuminate the ground surface in case of emergency landings at night.[5] The ground instruction program utilized lectures and films on a wide variety of subjects: for the aircrew, meteorology, R/T procedure, Oboe, Ground Controlled Approach (GCA) radio range, Morse, advanced formation flying, four-inch training flares, rocket projectiles, airmanship, and a host of topics having to do with jet aircraft, such as engines; Vampire airframes, performance and endurance; and Goblin engines – a sure sign of things to come. The airmen received instruction on aircraft types and their use, ground handling procedure, Vampire airframes, signals, and shop training.

"PAY ATTENTION!" FSGT FRANK KLAPONSKI POINTS OUT THE FEATURES OF THE DE HAVILLAND VAMPIRE'S GOBLIN II ENGINE

By 1 May 1947, it was decided at a general conference that the Squadron had reached the stage of development where the Auxiliary personnel could now assume a greater responsibility for its operation. Auxiliary groundcrew would, to a large extent, be responsible for maintenance of the aircraft. Over the years, the Squadron also engaged in two other standard types of activities – summer camps and war-game exercises, the latter involving other units, Auxiliary as well as Regular and, in some cases, elements of the USAF.

As a fighter-bomber formation, 402 Squadron was part of Canada's tactical air force, charged with aerial support of land forces. As far back as 1945, the Minister of National Defence had announced the creation of the "Mobile Striking Force" (MSF), an Army brigade group intended to provide an immediate and rapid counter to enemy lodgements that might seek to establish a foothold on Canadian soil. The MSF was to comprise three airborne battalions and appropriate artillery and supporting elements (e.g., communications and medical units, and engineers to construct bridges, airfields, etc). To the RCAF would fall the responsibility of transporting paratroopers and infantry, towing troop-carrying gliders, evacuating casualties, and providing aerial reconnaissance and offensive support and interdiction. It was these latter functions to which 402 Squadron would contribute with its fighter-bombers.

It should be borne in mind that for several years following the war, the Auxiliary was Canada's first line of defence insofar as the RCAF was concerned, providing the bulk of domestic air defence forces until 1955. The mandate of the Auxiliary was "to provide a first line reserve of fully organized, manned and equipped squadrons which can be mobilized on short notice into fighter and tactical components."[6] The first postwar Regular Force combat squadron, No.410, did not form until 1 December 1948; until then, only the RCN (interestingly enough) and the RCAF Auxiliary possessed operational squadrons, and 402 was in the vanguard.

CHAPTER 3

The North West Air Command Years
(March 1947 - August 1951)

In a reorganization of air commands to conform to post-war plans, Eastern Air Command, Western Air Command and No.2 Air Command were disbanded effective 1 March 1947. Two geographical commands were now in place: the newly formed Central Air Command with headquarters at Trenton, and the already-existing North West Air Command, headquartered in Edmonton. The over-riding responsibility of the latter was the defence of northwestern Canada in conjunction with the Army's Prairie and Western Commands. March 1 also witnessed the creation in Winnipeg of a satellite headquarters, No. 11 Group, of which 402 now formed a part, under the control of North West Air Command. The Group was charged with developing air support doctrines and techniques appropriate to the defence of its stated region of responsibility. Finally, on this same date, 402 Squadron, still armed with the Harvard, converted to a fighter role as No.402 (F) Sqn (Aux).

MARK 4 HARVARDS WERE NOT COMMON IN THE RESERVE SQUADRONS. AT ONE POINT, WHILE IN 402 SERVICE, THIS AIRCRAFT WAS FITTED WITH A GYROSCOPIC GUN SITE. NO DOUBT A RESULT OF ITS PREVIOUS LIFE AT RCAF STATION MCDONALD, A GUNNERY SCHOOL

HARVARDS, FEATURING THE DISTINCTIVE 402 SQUADRON BLUE COWLINGS, ON THE LINE AT WINNIPEG

The decision to assume a fighter interceptor role must be viewed in light of political and military developments taking place elsewhere in the world. In late 1948, the Communists executed a *coup d'état* in Czechoslovakia, much to the alarm of the West. This incident is regarded by many historians as the beginning of the "Cold War". However even before this, events had given cause for concern: as far back as July of 1944, a USAF B-29 had landed at an airfield at Vladivostok Bay due to fuel shortage. Later that year, similar circumstances forced two more B-29s to land in Soviet territory. These aircraft were impounded by the Soviets and served as pattern machines for the production of their own version of the B-29 under the designation TU-4. Three such aircraft appeared at the Soviet Aviation Day Display at Tushino aerodrome in 1947. This news was greeted with considerable misgivings by the Western powers, as the Soviets now had a long-range strategic bomber (some 200 were estimated to have been in existence in 1948) capable of delivering attacks on North American cities. The sole means of dealing with such a threat in the late 1940s was the manned interceptor.

Rather more mundane were developments in the marking of Squadron aircraft pursuant to Air Force Routine Order 250/47 that went into effect in May 1947. When initially received, the Harvards were still wearing their wartime BCATP livery. In keeping with the new International Civil Aviation Organization (ICAO) system of individual aircraft registration, all RCAF machines were now to carry the two-letter national service designator "VC", plus a two-letter squadron code followed by a single-letter individual aircraft identifier allocated by the squadron. No.402 squadron's designator was "AC". As of late November of 1947, Support personnel were engaged in repainting the Squadron's aircraft with the new letter call signs.

HARVARDS ARMED WITH ROCKET PROJECTILES (RPS) HEAD FOR GIMLI

By the second quarter of 1947, it had been determined that 402's operational aircraft would be the de Havilland Vampire. Was the Squadron looking forward to flying jets?

Some hint as to the level of anticipation can be gained from the OC Support's monthly report of proceedings for May in which he noted that "F/O Hardy returned from the Vampire course at de Havilland Aircraft Co.'s plant in Toronto with 'pukka gen' oozing out of every pore."[7]

1948

Shortly after the cessation of hostilities in 1945, a two-year postwar "Interim Period" had been instituted, the focus of which was to be the demobilization of some 90% of the wartime force. The Interim Period ended on 30 September 1947, and Air Force personnel who had agreed to serve during that time were released. Simultaneously, the RCAF, which had been on active service since 1 September 1939, stood down and an intensive recruiting campaign commenced. Air shows, already a familiar scenario to 402 Squadron, became a high priority in an effort to stimulate public interest and to encourage civilians to enlist in military careers. To this end, Squadron Harvards took part in air shows at Brandon and Yorkton in late May of 1948. Home base facilities saw continued improvement with the installation and operation of air-to-ground communications equipment in 3 Hangar by the Signals Section in February.

April 19th 1948 was a banner day for the Winnipeg Bears; in the afternoon, a large number of station and Squadron personnel gathered at 3 Hangar for the expected arrival of 402's first Vampire jet fighter. At around 1600 hrs, the

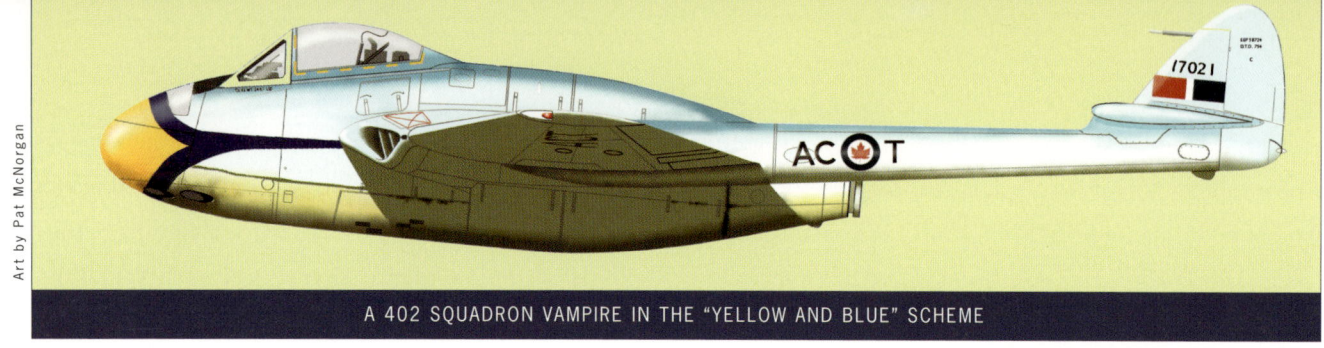

A 402 SQUADRON VAMPIRE IN THE "YELLOW AND BLUE" SCHEME

little jet whistled into view and did several beat-ups of the aerodrome before landing and taxiing to the hangar. 402 Squadron had entered the jet age.

A VAMPIRE COMES IN FOR A LOW PASS OVER THE FLIGHT LINE. THE AIRCRAFT HANDLED MUCH THE SAME AS THE HARVARD, AND WAS WELL LIKED BY THE PILOTS. UNFORTUNATELY, IT DID NOT TAKE WELL TO THE HARSH CANADIAN WINTERS

The Vampire's arrival created another opportunity for a public relations event, so on the 24th of the month it was placed on display. Some 3,000 people visited No. 3 Hangar to look it over and ask questions of Squadron personnel on hand for the occasion. The static display was followed by a flying demonstration, which was to include the jet forming on a flight of six Harvards. This proved to be a challenge because even with its dive brakes deployed, the Vampire passed the Harvards as though they were standing still. Air Force Day was celebrated with ground exhibits and an air demonstration at Station Winnipeg on 12 June, featuring Squadron Harvards and the Vampire, and 417 Squadron Mustangs from the Joint Air School (JAS) at Rivers.

A VAMPIRE ON THE RAMP AT WINNIPEG

During the first half of 1948, an AMES (Air Ministry Experimental System) 11C mobile radar convoy was activated at 11 Group: on 11 April the Squadron's Signals Section had the diesel-power unit and scanner in operation for the first time, and by 20 May it had been set up at the end of the tarmac near No.2 Hangar, with calibration runs being carried out by Squadron aircraft. Things were now in place for Squadron pilots to conduct interception exercises plotted by the radar convoy. These were among the first steps toward a separate and more formalized aircraft control capability that would come into being two years hence. In the meantime, the unit conducted a number of interception exercises in conjunction with 402. As these advances took place, yet another inaugural event was in the offing: the Squadron's first Army co-op exercise was conducted in the Birds Hill area north of Winnipeg on 20 June, when seven Harvards and the Vampire conducted simulated bombing and strafing runs.

This activity invites analytical comment. Although 402 had now been officially designated a fighter squadron and its operational aircraft type was an interceptor, the fighter-bomber/ground support role was continued. The Vampire's weaponry was relegated to four cannons, but these afforded a strafing capability. The Squadron's Mk IIA Harvards, each armed with a .303 machine guns and practice bombs, were also used in air-to-ground armament exercises.

A HARVARD START. THE GROUNDCREW HAD TO PAY PARTICULAR ATTENTION TO THE PROPELLER, LOCATED AS CLOSE AS IT WAS TO THE EXTERNAL POWER RECEPTACLE

Meanwhile, conversion on the jet was underway, and three more machines (17023, 17032, 17038) arrived on 25 June, giving the Squadron a total of four jet aircraft to date.[8]

More public events were staged, including the local Optimist Air Show in early July. This one was of interest in that it featured, along with 402's Harvards and Vampires and three Mustangs from JAS Rivers, a pair of Mustangs from the Minnesota Air National Guard. This favour by the Americans was returned in spades later that summer, when the four Vampires took part in the 1948 Minnesota State Fair. Before that came about, however, the Squadron devoted a fortnight in early July to its inaugural postwar summer camp at RCAF Station Gimli, during which time it accumulated a total of 327 flying hours in the seven Harvards and three Vampires deployed for the purpose. At the top of the agenda was jet conversion: three of the pilots had already been checked out on the Vampire ahead of time, and during camp, ten more joined the ranks of the "Untouchables," as the jet pilots were called. Bombing practice at the Shilo range and interception exercises were also major components of the summer camp which, upon its completion, was generally agreed to have been a huge success. On the 24th of the month, the Squadron's fifth Vampire was delivered in good order, and five days later the Squadron held a smoker for all ranks in appreciation for the good work done by 402 Support groundcrew at the recent summer camp.

CHAPTER 3

THE "UNTOUCHABLES"; THE NICKNAME, TAKEN FROM THE POPULAR TV SERIES OF THE TIME STARRING ROBERT STACK, APPLIED TO THE PILOTS. UNTOUCHABLE BECAUSE OF THE GREAT SPEEDS (MAX 540 M.P.H.) THEY WERE ACHIEVING WITH THEIR NEW JET VAMPIRES. ON THE LEFT IS W.R. JAMES, FUTURE CO. D.M. GRAY IS IN THE CENTRE, AND VERNON BASTABLE STANDS AT THE FAR RIGHT

JUST OUTSIDE OF 3 HANGAR, WINNIPEG, W/C LORNE CAMERON READIES FOR FLIGHT IN A VAMPIRE BEARING THE 'CITY OF WINNIPEG' CREST

We now return to the topic of the Minneapolis State Fair, where the Squadron's Vampire jets put on a demonstration of aerobatics and low-level formation flying over a ten-day period beginning on 28 August. The significance of the event was not lost on Winnipeg's City Fathers, who provided $200.00 to have the City crest painted on the noses of the Vampires. The Canadian aircraft were the only ones at the fair, which attracted a million people from all over the United States. The Squadron and North West Air Command proved so popular that a special "Canada Day" was declared. The huge crowd roared its approval when A/V/M K.M. Guthrie, Air Command, was introduced. Pilots and their groundcrew were also able to compare notes with their counterparts in the Minnesota Air National Guard fighter squadron, who also trained on a part-time basis while holding civilian jobs.

Unlike the Canadian Squadron, the American National Guard was not equipped with jets, and intense interest was shown in the Vampires from Winnipeg. S/L Tommy Anderson, S/L Jack Hudson, F/L Jack Dempster and F/L Matt Reid attracted so much attention in their manoeuvres before the grandstand – which on one occasion held 36,000 people – that they almost turned into media personalities. The pilots appeared on thirteen radio programs of fifteen-minute duration, and on two television programs. With an eye to further cementing the friendly ties between Canada and the United States, preliminary arrangements were made for exchange visits between the RCAF Auxiliary personnel and the Minnesota National Guard squadron. Officers of 402 hoped that an annual exchange flight might be made to encourage sharing of training and defence ideas.

In late September, several of the Squadron Harvards participated in a three-day searchlight co-op exercise[9] over Winnipeg to help publicize Army Week. This type of exercise was destined to become an annual event in future years. Further involvement with the Army came on 26 September with the staging of the Tactical Brigade Group's "Exercise Buffalo" in the Pine Ridge area north of the city. The Squadron provided air support, and simulated strafing attacks and tactical reconnaissance with the use of three

Vampires and seven Harvards. Practice bombing attacks were carried out on Army emplacements using flour bombs.

Two important events marked October 1948: on the first of the month, an RCAF (Auxiliary) University Flight was formed in the Squadron. To stimulate interest in the Air Force and to provide a flow of trained university students as commissioned officers for the Regular, Auxiliary or Reserve components, RCAF (Auxiliary) University Flights were organized across the country commencing in August of 1948. Each Flight had an establishment of one hundred cadets, with selection being made at the rate of approximately thirty-five freshmen annually. The training program covered three years. Lectures were given during the school year, and the undergraduates were paid as Flight Cadets while receiving flying or specialist training during summer vacation. Across the system, the summer training covered three aircrew and eleven groundcrew categories. The University of Manitoba's (U of M) Flight came into being on 1 October 1948, and was among the first such units formed. The Squadron on the air station provided accommodation and training facilities until suitable quarters were ready on the university campus, and two U of M Flight Cadets received refresher flying training during November. The second important event involved equipment. Three flights of the Squadron's nine Harvards were flown to 11 Technical Servicing Unit (TSU) Montreal over the course of October and early November. There, they were exchanged for overhauled and modified aircraft with upgraded radio equipment including radio compasses.[10]

At the end of October, there was an investiture at which four of the Squadron's members received decorations. Two of the awards were Military Crosses (MCs), conferred upon S/L R.G. Johnson and F/O V.J. Bastable in recognition of their wartime accomplishments. As only five MCs were awarded to RCAF personnel pursuant to their combat experiences, 402 was proud to have two of the recipients on its rolls. The two other airmen, F/Os L.M. Cameron and H.N. Scott, received Distinguished Flying Crosses (DFCs).

W/C Clement's tour as squadron CO came to a close on 21 December and the following day, the duties of command were taken up by W/C L.M. Cameron, DFC, who ascended three ranks upon his appointment. Like his predecessor, Lorne Cameron was a former 402 pilot, having served with the Winnipeg Bears overseas during the war. The year was closed out with extensive tactical, formation, night flying and interception training in the Harvards during December. The Vampires were largely grounded during this time pending modification of their oleo legs.

S/L R.G. JOHNSON, (LEFT) AND F/O V.J. BASTABLE ON THE OCCASION OF BOTH OFFICERS BEING AWARDED THE MILITARY CROSS

To train new pilots for the Auxiliary, in December 1948, the Air Force introduced a scheme whereby Auxiliary squadron commanders could sponsor direct entry applicants or serving Auxiliary airmen who possessed the same qualifications as were required for aircrew training in the Regular Force. Cadets would be appointed as Flight Cadets in the Auxiliary for aircrew instruction and, on graduation with "Wings" standard, they would be commissioned in the Auxiliary and would return for non-continuous service to the squadron that promoted them. Initially, the scheme provided for an entry of twenty selected candidates on a quota of three per squadron. Subsequently, provision was made for additional Auxiliary selections to fill any vacancies in Regular aircrew training courses. A similar scheme was introduced in January 1949 for the training of Auxiliary groundcrew. Direct entry candidates requiring complete trade training, or serving Auxiliary airmen who required higher trade proficiency, could be nominated for vacancies in Regular trade training courses.[11]

1949

For the Winnipeg Bears in particular, the first quarter of 1949 was marked by weapons exercises with the Harvards and their .303 machine guns at the Shilo range, made noteworthy by the Squadron's first use of rocket projectiles in March. Also during March, the Vampires were returned to service after an extended period of unserviceability due to the needed modifications to their oleo legs. Tragically, the Squadron suffered its first flying fatality on 27 March 1949 when F/O Bastable, MC, was killed during a routine

THE CRASH SITE OF F/O BASTABLE'S VAMPIRE IN THE WINNIPEG SUBURB OF CHARLESWOOD

training flight in his Vampire. He was posthumously promoted to Flight Lieutenant and is remembered by the Squadron through the Vernon Bastable Memorial Award, instituted in 1993 and given to 402's Regular Force "Airman/Airwoman of the Year."

In the spring of 1949, the Squadron again held forth with flying displays, this time sending a pair of Vampires to air shows and displays at Regina, Moose Jaw, and Wilcox, Saskatchewan. The Squadron reverted to its former fighter-bomber designation on 1 April as "No. 402 (FB) Sqn (Aux)." The same day, the parent No.11 Group was re-designated "Tactical Group," comprising 402 (Winnipeg), 403 (Calgary), 406 (Saskatoon), and 418 (Edmonton) squadrons. To Tactical Group fell the responsibility, in collaboration with Air Defence Group and Air Transport Command, of training aircrew in offensive and transport support.[12] Three additional armament Harvards were delivered, and several training exercises, using both the Harvards and the Vampires, were carried out in co-operation with the Army. One of these, the Reserve Army's "Operation Chestnut II", involved an air contact team and four Harvards providing cab rank support[13] with the dreaded flour bombs. In early June, two Regular Force officers from the Canadian Joint Air Training Centre (CJATC) (formerly the Joint Air School) arrived at the Squadron for Vampire conversion training, but the highlights of the month comprised publicity programs, one of which was especially novel.

Between the 6th and the 11th, Squadron aircraft helped celebrate the City of Winnipeg's 75th anniversary. Honourary command of the Squadron was handed over to the mayor for the week of celebration, and every day at noon, Vampires flew over the main business section. Radio contact with the aircraft was set up on a stand at the corner of Portage and Main, and His Worship gave flying instructions to the pilots of the aircraft flying over the intersection. The stand also had a broadcasting hook-up. Six Harvards conducted early-evening formation flying over the city. Later in the month, Squadron aircraft put in additional public appearances at Gimli and Fort William, Ontario.

Between 4 and 17 July, the Squadron again deployed to RCAF Station Gimli for summer camp. In addition to 402 Squadron with their Vampires and mobile radar convoy, this year's participants included 442 (City of Vancouver) Auxiliary Squadron with their Vampires, Harvards and Expeditor, and Mustangs from 403 (City of Calgary) and 417 from CJATC Rivers. Numbers 406 (City of Saskatoon) and 418 (City of Edmonton) squadrons provided the bomber force armed with B-25s. Following a warm-up phase for the mock battles to follow, Gimli was declared to have been captured and all personnel based at the station were designated the "enemy." To put matters right again, a glider train from the CJATC arrived with an Army assault force. A wireless team quickly set up shop and was soon directing an air strike in support of the troops attempting to recapture the airfield. The air assault was spearheaded by the 406 and 418 Mitchells roaring over the enemy flight line with bomb bay doors open and props at fine pitch. Bombs of brightly-coloured flour sacks tumbled out, splattering among the parked aircraft, engulfing the tower in clouds of flour and smashing into the tower walls. Some groundcrew were hit as they ran to re-arm and refuel their aircraft, but thankfully the sacks caused no serious injuries despite slamming into the hapless airmen at 180 mph.

A 406 SQUADRON B-25 MITCHELL. 402'S EXERCISES FREQUENTLY INVOLVED THESE AIRCRAFT

Revenge was swift and vicious once the enemy Harvards and Mustangs got airborne. The bombers were not left defenceless, however, and their fighter escorts plunged to the attack. By 0900, yellow-nosed aircraft were fighting it out with red- and blue-nosed opponents directly over the heads of the enthralled observers and umpires. In 25 minutes, the re-armed Mitchells were again diving into the fray, at times flying so low that newsmen in the control tower had to shoot downwards to get aircraft in the picture.

After the initial softening-up, the air cover formed cab rank patrol in twos over their own territory until called upon by the tactical controller. At the same time, the air was filled with whining crescendos as Vampires were being scrambled by the radar controllers to meet the next wave of attackers. The umpires – 30 in the air and 15 on the ground – scored the fighters' attacks. The pilots were assessed on their use of the sun for the bounce and the proficiency of "curve of pursuit" for the gunnery. Different forms of attack were worth different scores, and to be credited with a kill a pilot had to be given an aggregate score of eight points.[14] Such was the colour and drama experienced by 402 Squadron at its summer camps when it functioned as a front-line combat unit in the early years of the Cold War.

During the first week in August of 1949, eight 402 Squadron Harvards participated in "Exercise Eagle" in the Fort St. John-Grande Prairie area. Planned and executed jointly by the RCAF's North West Air Command and the Army's Western Command, Exercise Eagle sought to exercise the PPCLI in its role as an airborne/air-transported infantry battalion; to exercise RCAF Auxiliary fighter and light bomber squadrons in offensive support, and Regular transport squadrons in support of airborne/air-transported operations; to give Army/Air staffs experience in the preparation and conduct of joint operations; and to familiarize and exercise RCAF staffs in joint planning and logistic problems in support of tactical operations in northwestern Canada. This statement of goals exemplifies rather well the mandate and structure of that part of the Canadian military establishment to which 402 belonged at the time. To the surprise of no one, the flying of the permanent force proved to be superior to that of the Auxiliaries. Especially in breakaway following ground strafing and in curves of pursuit during air combat, the Regular pilots possessed a noticeable edge, even bearing in mind the greater speed of their Mustangs relative to the Auxiliaries' Harvards. Nonetheless, the experience was instructive in defining for the "Weekend Warriors" the standard to which they would have to aspire to become effective combat pilots on par with their Regular Force colleagues.

Four Vampires once again took part in the Minnesota State Fair between 23 August and 5 September. Two days before heading south, one of the aircraft blew a tire on take-off, seriously damaging the undercarriage. To ensure that the fair-goers were treated to the full complement of four aircraft, the injured bird was replaced with another of its kind flown in from the sister Auxiliary Squadron based at Vancouver. On 26 August, 402 established a new Winnipeg-to-Minneapolis speed record, completing the trip in a tidy sixty-four minutes at a cruising speed of 410 mph.[15] Effective 15 September, the RCAF (Reserve) University of Manitoba Flight was struck from the establishment of 402 (Res) Squadron and placed under the functional aegis of

North West Air Command and administrative control of Station Winnipeg. The month of September was rounded out with an air display over the town of Selkirk in support of a local Air Force recruiting drive.

To be realistic, it may not be possible to count on the amenities of home should hostilities break out: it may in fact prove necessary to operate out of a base in which facilities are minimal to non-existent. Accordingly, the Air Force training "syllabus" included mobility exercises that involved moving to alternate locations and operating from there for a time. One such temporary change of scene came early on Sunday, 6 October, when the Squadron moved en masse to the abandoned Portage la Prairie aerodrome and conducted the day's flying ("Operation Portage") from there. The Winnipeg Bears' nine Harvards were followed within minutes of their arrival by ground personnel – fitters, riggers, armourers and cooks – embarked in a Dakota and an Expeditor. Twenty minutes after that, six Harvards, armed with flour bombs, were off on a tactical exercise to the south.

As aircrews came in from their sorties, they lined up for an open-air meal before being briefed by the Intelligence Officer for another bombing flight. Technically, the Portage airfield was knocked out when the first six aircraft became an attacking force on their return journey. A flight took off to defend the aerodrome, but the "enemy" slid in under the defenders and "bombed" the dispersal point. Interceptions proved better in the afternoon: approaching Pembina Lake, the Beechcraft, with members of the press on board, was jumped by four Harvards returning from the Shilo bombing range. That formation immediately claimed a victory. Shortly afterwards, a second group of Harvards made quick work of the Expeditor again. Undaunted, the Beechcraft returned to Portage, only to find that the first four Harvards were waiting once more over the base and the woebegone "Bugsmasher"[16] bit the dust for the third time that day. Nor was the Expeditor the only victim of the murderous Harvards – a signals convoy of three trucks sent out from Winnipeg was also done in.[17] On 6 November, another mobility exercise was carried out, this one from the Brandon airfield.

Meanwhile, the world situation, from a Western standpoint at least, continued to deteriorate: by early August 1948, the Soviets had blockaded Berlin, a move which precipitated the now famous Berlin Airlift. Of even greater concern was their detonation of an atomic bomb in 1949. The Russians now had both the bomb and an aircraft capable of delivering it anywhere on the North American continent. As the decade of the '40s drew to a close, fighter defences were becoming an ever-increasing matter of importance in Canada and the United States.

1950

For 402 Squadron, the year 1950 was one marked both by continuity and by change on several fronts. Extensive tactical, formation and interception flying training, homing exercises, air combat, cine gun, and air and ground firing exercises were carried out; and over the course of the year flying demos were laid on at the Kenora Snow Carnival, the Stonewall Kinsmen Carnival, the Saskatoon Air Force Day, the Rivers Armed Forces Day, the Beausejour Dominion Day celebrations, and the Calgary Air Show. W/C Cameron, DFC, resigned as CO, and on 28 February, command was taken up on a temporary basis by W/C B. Breckon, DFC, who was promoted to the rank of Squadron Leader and on the authority of AFHQ, assumed command of 402 Reserve Squadron on 1 May. His first task was overseeing the unit's contribution to fighting the massive flood that struck Winnipeg in May. The entire Squadron was called to duty in the emergency, and assumed responsibility for a section of the dyke behind St. Boniface Hospital. Both aircrew and groundcrew worked shoulder-to-shoulder, piling up sandbags and patrolling on a 24-hour basis.

HARVARDS LINED UP FOR "FAMILY DAY" OUTSIDE NO.3 HANGAR

Nor was 1950 without its flying accidents: a particularly bad one occurred on 16 April when F/O V.E. Barber, on his third flight in a Vampire, crash-landed in a garden at the rear of a College Avenue residence south of the airfield. Tragically, a civilian, working in his garage at the time, was killed, while F/O Barber sustained serious injury and was admitted to nearby Deer Lodge Hospital. Additional Vampires were ferried in during May to replace those lost in the recent accidents. Poor-quality radio transmission and reception of the transceiver was addressed by the installation of a VHF set in the tower at No.3 Hangar in June. Numerous operational exercises were carried out as well, i.e., interceptions, air combat and cine gun, with the aid of the radar teams. Various homing exercises came within 2- to 3-mile accuracy, and several very successful one-station fixes were also obtained.

Before the annual summer camp got underway, air-to-air firing trials were carried out by a Squadron Harvard and a Vampire at the Lake Manitoba range against a banner target towed by Mustangs from Rivers, and air-to-ground firing trials against a gunnery screen target at a range near Sleeve Lake northwest of Gimli, also with a Harvard and a Vampire. The purpose of these late June trials was to establish the best procedures for future armament exercises. For the air-to-air work, each Squadron aircraft flew a single sortie, and coloured ammunition (blue for the Vampire's cannon shells, orange for the Harvard's .303 machine gun bullets) was used to distinguish between hits on the targets. This method of using painted ammunition left a lot to be desired because the dye did not always come off and make the intended mark,. In addition to the machine gun ammunition, the Harvard was also armed with four 3" rocket projectiles and eight $11^{1}/_{2}$-lb practice bombs for the air-to-ground work. The 1950 Gimli-based summer camp then commenced with general flying and air-to-air/air-to-ground armament programs at the Lake Manitoba and Fisherton Lake ranges, followed by tactical exercises. Local media coverage of the operational phase of the 1950 summer camp read as follows:

> Mythical "enemy" forces were swept from the skies Monday near Gimli, Man., as they moved in with an "invasion force" from Hudson Bay.
>
> The five reserve squadrons of the R.C.A.F. in western Canada were reproducing war conditions as closely as possible at the beginning of their second week of training at Gimli.
>
> The City of Winnipeg's 402 squadron took the role of an enemy force trying to wipe out air resistance on the route to Winnipeg.
>
> Making a "successful interception" were the squadrons from Calgary and Vancouver. Meanwhile, air squadrons from Edmonton and Saskatoon blasted enemy installations at the airport at Kenora, Ont.
>
> As soon as radar units at Netley and Gimli had picked up the approaching enemy forces, they relayed the information to operation headquarters at the Gimli R.C.A.F. station. Within 38 seconds of

A 402 VAMPIRE CREW STOPS WORK TO ADMIRE ONE OF THE GREATS: A B-25 MITCHELL

> the firing of a "Very's" [sic] pistol, the Calgary squadron had two sections (two Harvard aircraft each) in the air and under headquarters control by radio.
>
> The Calgary squadron, acting as a tactical wing, spotted the enemy aircraft (Winnipeg's squadron) and moved in to intercept them. Veering off, the enemy came in for a second attack about half an hour later, only to be demolished by the Vancouver squadron.
>
> In the afternoon, there was a similar result with other personnel of the same squadrons. It was a realistic battle with the "home" forces having air superiority. Both sides were using Harvard aircraft. In the meantime, the enemy had been successful in taking over the Kenora airstrip and Fort Frances, Ont., as well as isolated points to the north.

The Edmonton and Saskatoon medium bomber squadrons, flying Mitchells (B-25's) played havoc with enemy positions at the Kenora airport in two separate raids. This left the way open for the First Canadian Army to move in heavy equipment along Highway No. 1 to recapture the airport.

All this activity was just the opening day of the second week of training for the western reserve squadrons. And the day opened at 5 a.m. After breakfast there was a 6 a.m. briefing under Group Captain M.P. Martyn, senior air staff officer, north west air command, Edmonton.

His officers pointed out to the various squadrons and to 11 members of the press and radio of western Canada, details of the day's operation.[18] Nor was Gimli itself free from counter-attack as the force from 402 Squadron swooped down on the "first-time" groundcrews with their stinging load of flour bombs.[19]

The annual arrival of the Central Flying School Visiting Flight in late September 1950 resulted in the majority of the 402 pilots receiving their proficiency checks. On the 18th the Squadron's title was officially amended to "No. 402 'City of Winnipeg' (FB) Sqn (Aux)," indicating that a new role was in the offing for the Winnipeg Bears. By this time, a decision had been made to replace the Vampire with the North American P-51D (Mk IV) Mustang tactical fighter as the Squadron's operational aircraft type. Part of the reason was logistical: replacement parts were difficult to acquire because the Vampires were British aircraft. The RAF was also flying Vampires; they needed parts too, and the colonials had to wait their turn. It was reaching the point where aircraft had to be cannibalized for scarce parts. Beyond that, the Vampire Mk III, originally designed as an interceptor, was not well-suited to the fighter-bomber role with which the Squadron was now tasked. At low level with throttle wide open, it had a very short 55-minute endurance in the ground support role. For the latter, it had only cannon to bring to bear; there was no provision for bombs or rocket projectiles.

AN EARLY 402 MUSTANG PATROL. THE CLOSEST AIRCRAFT IS NOT YET PAINTED IN NATIONAL MARKINGS

The Mustang, on the other hand, was armed with six .50 calibre machine guns and was capable of carrying ten 5-inch aerial rockets or 2,000 pounds of high-explosive or napalm bombs. The Mustang could stay up for seven to nine and a half hours, as opposed to two hours for the Vampire.[20] While they were of Second World War vintage, it must be remembered that Mustangs were being used to good effect as fighter-bombers by no fewer than four air forces in the Korean conflict.[21] The loss of the jet-powered Vampires and their replacement by piston-engine aircraft of Second World War vintage was compensated for by the fact that more operational aircraft would now be on strength. In preparation for the arrival of the Mustangs, Squadron officers proceeded to CJATC Rivers in October to attend Offensive Support and Mustang Conversion courses with Support groundcrew personnel taking Mustang courses there as well. Also in October, the Squadron consolidated itself in No.3 Hangar; and the yearly staging of "Exercise Assiniboine," a tri-service demonstration of air power in support of an army staged for the benefit of students of the Canadian Army Staff College, came and went on 14 November.

Of major significance was the introduction in 1950 of a Wing-level organization within the RCAF, which gave reserve personnel a greater share in the administration and control of their own activities. Additional to the flying squadrons, the Auxiliary created specialized units to back up those squadrons. Wing headquarters, established in urban centres where two or more such Auxiliary formations were located, would provide administration for an aircraft squadron, an aircraft control and warning unit, an intelligence unit, a technical training unit, and a medical unit. On 1 October 1950, the Air Force Auxiliary was denominated the "Reserve,"[22] and "RCAF (Reserve) Wing Winnipeg," reporting to North West Air Command through Tactical Group, was formed on that date to administer and control 402 Squadron and a number of support units, notably 3052 Technical Training Unit, 2402 Aircraft Control and Warning Unit, and 5001 Intelligence Unit that had been created to back up the Reserve flying units and the Regular Force. In October of 1949, a medical unit was established in Hamilton as part of the Auxiliary formation there. By March 1951, fifteen additional such units had been set up across the country, including 4003 Maintenence Unit (MU) in Winnipeg.

Of particular interest is 2402 Aircraft Control & Warning (Reserve) Unit, which came into being at Winnipeg on 1 October 1950 under the administrative control of the local

Reserve Wing headquarters reporting to North West Air Command through Tactical Group. Most of the 402 officers and NCOs of the Signals Trade were transferred to the new unit, whose equipment continued to be Second World War-vintage AMES 11C radar mounted on four trucks along with similar-vintage radio gear in two more trucks. At the outset, S/L Fred D. Searles, an officer of wide experience in the design, production and operational use of VHF radio equipment, commanded 2402 AC&WU. High priority was given to obtaining the services of former radar technicians and operators for the new organization, but even more emphasis was placed on bringing in younger men wishing to obtain training in this field. The Unit was equivalent in status to Reserve squadrons already in existence, with equipment, pay and administrative officers of its own. Weekend commitments involved round-the-clock radar watch over Manitoba's skies. It was the type of unit from which, it was expected, civil defence organizations would receive their air-raid warnings for transmission to the public. Besides keeping watch on air traffic in central Manitoba, the new unit formed an integral part of the high-speed, high-altitude interception syllabus carried out by 402 Squadron. Its use in Western Canada contributed greatly to the efficiency of radar control and warning in Winnipeg. Although an independent unit in its own right, 2402 AC&WU attended summer camp with 402 Squadron on several occasions in the early 1950s. The main role of the Reserve AC&Ws was to provide trained manpower in event of mobilization for war. The Reservists would allow the Regular AC&W squadrons to operate on a 24/7 basis, something the Regular Force units did not have the manpower for during peacetime.

2402 OPERATIONS TABLE

No.5001 Intelligence Unit comprised fifty-eight personnel at the time of its formation and was made up mainly of ex-aircrew and ex-fighter pilots. Their prime objective was to brief and de-brief aircrew and to evaluate information on the basis of photographic interpretation and crew reports. Its first Commanding Officer was F/L A.S.R. Tweedie who had been intelligence officer for 402 Squadron. The newly-formed technical training unit, 3052 TTU, provided instruction in aero-engines, airframes, instruments, electrical systems, munitions and weapons, and mobile and safety equipment. Located in Building 16, this unit was commanded by S/L W.R. Lee when it was first set up in 1950. The staff comprised Auxiliary officers and airmen supplemented by a number of Regular Force personnel and a few civilian instructors. Trade advancement for Auxiliary technical tradesmen who did not hold the highest outright grouping for their employed trade was one job faced by the instructors of the TTU.

ENGINE INSTRUCTION AT THE TTU

On 24 November, the first of 12 Mustangs destined for the Winnipeg squadron arrived from Burbank, California. In that connection, a letter was received from Tactical Group on 12 December directing that there was to be no more flying in Vampires by 402; for the time being, flying was to be limited to Harvards. The Squadron's first Mustang was flown on 29 December 1950. In late November, meanwhile, tactical exercises were carried out with 6 Division RCASC involving an air contact team. In this set-up, a VHF set installed in a jeep was operated by one of the pilots to control simulated air attacks on the RCASC column.

A MUSTANG IN THE FIRST OF TWO 402 COLOUR SCHEMES

1951

Good progress was made in January 1951 on Mustang conversion, as 19 pilots were checked out on type during the month. These proceedings were not without their drawbacks, however; the first P-51 prang transpired when one of them crash-landed at the end of the runway due to engine failure. Fortunately, the pilot was uninjured. Yet another incident, this one more humorous than serious, unfolded one day in March. During take-off, a Mustang's canopy blew open, hitting the pilot "a lusty clout" on the forehead as the coupe slid backwards. When power was reduced to the climbing setting, the canopy blew shut, again hitting the pilot – on the *back* of the head; this in the days before the advent of hard-hat type crash helmets! The trick to avoiding such experiences, of course, was to ensure that the canopy was locked shut on take-off. Of his introduction to the Mustang, Tom Patterson remembers:

> In preparation for conversion to the Mustang, the 402 pilots already flew Harvards and were proficient on them. We took the ground school, wrote the exam, then signed out a Mustang, started the engine and took off. I made my first flight in a Mustang on 10 January 1951 in 9600.

CHAPTER 3

We started Mustang training by flying individually until each pilot became proficient. The senior pilots would then match up with newly-converted pilots and tail chase through the sky, fly formation together, and various manoeuvres: to loops then slow rolls together until all were accomplished in these tasks; then down to low altitude until each pair were performing these tasks proficiently.

Once competent with two-aircraft formation flying, we went to four Mustangs together. And when adept with two sets we went to four sets, and finally formation flights with as many aircraft as could be mustered. Part of the training syllabus involved flying high-altitude, cross-country flights. When climbing to 35,000 feet, the supercharger cut in around 18,000 feet and up we went. At this altitude the aircraft handled a little sluggish, but really wasn't that bad. During the climb, the best check for oxygen anoxia involved the senses: this included checking the fingernails and looking in the rear-view mirror at your lips for cyanosis. We took the occasional decompression chamber checks and learned about oxygen difficulties.

As more and more Mustangs came on squadron strength, the fleet of Harvards was reduced. At one time we had upwards of 12 Mustangs.[23]

It is common for air squadrons to have mascots, and 402 certainly had theirs. He was a grizzly bear who hailed from near the town of Minto, in the Yukon. A resident

SGT MINTO IS PROMOTED TO FSGT. LEFT TO RIGHT: W.C. PACHOLKA, BILL SMALLWOOD, ED WOLKOWSKI

of Winnipeg's Assiniboine Park Zoo, he was formally appointed the Squadron mascot at a luncheon in the Royal Alexandra Hotel on 1 March 1951 and given the rank of Sergeant. His face became a model for the Squadron's crest, a replica of which was presented to the City and then placed on the bars of "Sgt Minto's" cage at the zoo.

In the spring of 1951, the Squadron finally divested itself of its Vampires, which were flown to Toronto on 5 April. The pilots who ferried the aircraft returned to Winnipeg the following day via Trans-Canada Airlines.

The first week in May brought with it "Operation Architect", a war-game production of 18 Wing in Edmonton. It involved a total of 25 aircraft from all four of the Group squadrons (six Mitchells each from 406 and 418, and five Mustangs from 402 and eight from 403). The tactical situation was as follows: an enemy force based at Churchill on Hudson Bay had advanced across Manitoba and Saskatchewan and was moving into eastern Alberta. Canadian and American units were proving successful in countering the intruders, but the friendly forces were meeting stiff resistance at a bridgehead across the Battle River at Wainwright. Allied units directly to the south were exposed to dangerous flank fire, impeding the Allied advance aimed at driving the enemy back into Hudson Bay. Close air support was called in to neutralize the enemy resistance at the Battle River, thereby allowing the allied army to cross to the opposite side with heavy artillery and armour, and enhancing prospects of renewing its advance in concert with the other allied forces. In this typical Korean War-era scenario, the Winnipeg Bears' responsibilities were twofold: bomber escort through hostile airspace, and air-to-ground attack.

"Operation Architect" was intended to be a learning experience, and under such circumstances shortcomings are as instructive as successes. The RCAF Station Namao-based fighter-bombers were a tad slow in getting off the ground, and 402 missed their rendezvous with the Mitchells for the intended escort phase. On the other hand, they and their Calgary counterparts received praise from Army observers for the growing accuracy with which they prosecuted their rocket and machine gun attacks on the target. All in all, notable strides had been made over what had transpired two years earlier in "Exercise Eagle."

On the last day of June, nine Mustangs and six Harvards departed for summer camp at RCAF Station Abbotsford. During this deployment, the Winnipeg-based Medical Unit, AC&WU, TTU and Intelligence Unit were also in attendance. The camp involved operational exercises in conjunction with the other Auxiliary squadrons under the direction of the aircraft control and warning units, and an armament program, somewhat curtailed by weather, at the Tofino weapons range. The summer camp was unfortunately conducted in the absence of F/O R.J. Dew, whose aircraft suffered engine failure after take-off back in May and crashed near the DOT control tower at Stevenson Field. The pilot later died of his injuries at Deer Lodge Hospital, becoming 402's first Mustang fatality.

The Tactical Air Group/Tactical Air Command Years (August 1951 - November 1953)

On 1 August 1951, North West Air Command was disbanded and absorbed by Tactical Group (formerly No.11 Group) to form "Tactical Air Group". Headquartered in Edmonton, its field of operation was expanded to include all of Canada. RCAF (Reserve) Wing Winnipeg became "No.17 (Reserve) Wing" on the same date, and exactly one month later the designation was again changed to "No.17 Wing (Auxiliary)." However these amendments did not alter the mandate or the activities of the Squadron. Three Harvards were transferred from 402 to Gimli and were ferried to that station on the last day of the month. A rare experience was enjoyed in October when Squadron officers lined the stairs of the Legislative Building during a Royal Visit to Winnipeg by Princess Elizabeth and Prince Philip.

A mobility exercise was carried out on 14 October along with 2402 AC&WU to test the units' proficiency in operating from a mobile or temporary base. It was originally intended that the exercise should move to and operate from Portage la Prairie, but the plan was changed at the last minute due to Portage being unserviceable. Armament work, navigation and interceptions were instead conducted at Gimli. On 8 November there was a Handing-Over Parade marking the retirement of W/C Breckon, DFC, from command of 402 due to civilian commitments. Command was assumed by S/L D.W. Rathwell, DFC, formerly a flight commander with the Squadron but more recently Assistant Operations Officer at Wing HQ.

By now, the Cold War was in high gear and the country's defence was front and centre on the national agenda. As shown in the 1951-52 White Paper on Defence, Auxiliary fighter and tactical squadrons and radar units were conspicuous in the overall scheme and a concerted push was on to enhance their operational standards. "Annual summer camps will be continued and increased emphasis will be placed on operational exercises with the regular force so that the minimal operational training will be required should it be necessary to place reserve units and personnel on active service in the event of a general war."[24] For 402 Squadron, the above-noted mobility exercise comprising armament, navigation and interception drills at Gimli preceded the 1951 rendition of the now-familiar "Exercise Assiniboine" that played out between 10 and 14 November at its usual Camp Shilo venue. The exercise simulated a friendly, amphibious assault force that became stalled while working its way inland from the coast. The obstacle was an enemy strong point that had to be neutralized by an airborne assault, with both land- and carrier-based aircraft providing close air support. Among the latter were Sea Furies and Avengers from far-off *HMCS Shearwater*, Nova Scotia, along with Mustangs from Rivers and a pair from 402 Squadron. Also during the month, the Air Force amended its aircraft marking system. The two-letter squadron code (in 402's case, "AC") was retained, but the individual aircraft identifier was replaced with the final three numerals of the particular machine's serial number.[25]

1952

The harsh conditions of a Manitoba winter formed the backdrop for "Exercise Strike" on 19 January 1952, when a dozen C-45 Expeditors, bomber stand-ins courtesy of No.2 Air Navigation School and flown by 406 and 418 crews (whose Mitchells were grounded), set course for the Selkirk Rolling Mills. The intruders were escorted by 403 Squadron Mustangs, and the thwarting of their mission was the responsibility of 402 Squadron under the control of 2402 AC&WU. The defenders scored a number of interceptions before the aggressors reached their target.

The Squadron's participation once again in "Exercise Assiniboine" in late February 1952 was followed by an extensive armament program at Shilo during the first week in March, which marked the Bears' first use of 500-lb high-explosive bombs. It should be noted that over the years they also conducted a fair bit of napalm bombing at Shilo. Each Mustang could carry a pair of 750-lb napalm bombs that were dropped from about 250 feet or so to produce a billowing ball of flame when they hit. One form of target especially suited to this type of ordnance was a static convoy of some 16 trucks laid out by the Army at the Shilo range. After releasing the bombs, the pilot could execute a clinching turn and watch the ensuing wave of flame engulf the convoy. Another target on the Shilo range was a derelict Lancaster airframe. When practicing Army co-op manoeuvres, the Mustangs would fly across the range and zoom in through their convoys, vibrating the antennas protruding from the vehicles. On occasion, the Harvards were "armed" with flour bombs and these were dropped on the troops below. Forward air control (FAC) exercises involved firing at targets specified by grid reference under the direction of a radio-equipped controller suitably positioned on the ground.

SOME CHRISTMAS CHEER AT HANGAR NO.4 WINNIPEG

The Squadron's singular contributions to the 1952 Air Force Day events at Station Winnipeg comprised a rocket attack on a target on the airfield and a low-level high-speed flypast in formation. Meanwhile, the 1951-52 White Paper on Defence called for "increased emphasis ... on operational exercises with the regular force so that the minimal operational training will be required should it be necessary to place reserve units and personnel on active service in the event of a general war."[26] To this end, the first three weeks of June were given over to extended navigation training in radio range and radio compass procedures before the Squadron repaired to summer camp and "Exercise Nugget" at far-away RCAF Station Watson Lake in the Yukon at the end of the month, again accompanied by Winnipeg-based 2402 AC&WU. Up until now, the summer camps were located in the southern and more populated areas of Western Canada. However, since it was essential that the Auxiliary squadrons of Tactical Air Group be given experience in flying over the northern areas of Western Canada, and operating from northern bases, the 1952 rendition of

A SCENE FROM WATSON LAKE

summer camp would be based at airfields – Watson Lake and Whitehorse – that formed part of the Northwest Staging Route in the Yukon Territory.

Altogether, some 400 Auxiliary personnel with 20 Mustangs attended the Watson Lake camp. For the first time in the history of these camps, women personnel were included. The idea was that the Wing Headquarters would plan and direct the camp with a minimum of Regular Force assistance and supervision. Broad operational objectives were issued to each Wing HQ, outlining fictitious military situations that could conceivably develop on the Northwest Staging Route and the Northwest Highway System. Each HQ was then responsible for planning and executing the necessary air operations to rectify these situations and assuming responsibility for carrying out routine training of units under its control. In addition to exercising Wing operations personnel in planning, executing, and controlling operations, the camps provided the Wing admin staffs with experience in planning and conducting summer camps. The mobility of the squadrons would be tested while they were gaining experience in flying over Northwestern Canada and their use of armament assessed. Each unit in the Wing would have the opportunity to train personnel in tactical roles, while the training staffs of Tactical Air Group HQ were available for specialist advice regarding such training programs.

F/O K. MANNS EXITING HIS MUSTANG AFTER COMPLETING ANOTHER OPERATION

For 402, the day of the big move came on 28 June, when 10 Mustangs and two Harvards departed Winnipeg on the first leg of the journey. All of the Mustang-equipped units – 402, 403 (Calgary) and 442 (Vancouver) squadrons – were based at Watson Lake. No.418 Squadron, with its Mitchells, operated out of Whitehorse, the enemy camp. The first day at Watson Lake was washed out by rain, and when the Squadron finally got down to its RP armament program and navigation trips, word was received that a Whitehorse-based

Mitchell of 418 Squadron had gone missing. Two Mustangs were dispatched in search of the Mitchell while the rest of the Squadron busied itself in air-to-ground rocketry and bombing, air-to-air firing, and navigation flights.

Bombing, gunnery and rocketry ranges, manned by members of the Auxiliary as well as the Regular forces, had been set up to meet operational training requirements. The dive-bombing and rocketry targets at Watson Lake were white triangles floating on the dark green background of the lake and sited where attacks were visible from the camp itself. The Winnipeg Bears' air-to-ground work yielded indifferent results at the outset, and the official explanation was that the Squadron was a bit rusty from not having conducted RP or bombing practice since March. Tom Patterson, one of the pilots, had another explanation:

> The target range at Watson Lake was located in the middle of the lake. They had an anchored floating raft set up for dive bombing using 500-lb bombs and rockets. Sighting positions were located on both sides of the lake that zeroed in on the raft to take readings; the observers told us how many yards we were off on our firing/bombing. I well remember making bombing runs on the target and the Range Officer kept telling me I was 10 yards, 15 yards off target. I contested his observations. They finally rechecked their equipment again and found that the target had moved. Apparently our first couple of rockets went down under the water and sheared off one of the anchor cables and the target was pivoting across the lake. Every time we fired our rockets, we hit the target but the Range Officer kept reading off the coordinates and reported our "error." I kept saying the rocket passed straight through the target. We carried live warheads on the rockets at the time.[27]

In any event, both the air-to-air and air-to-ground results improved as the camp progressed. Familiarization flights were flown for the airmen, and a very successful ground support exercise was carried out. On 8 July, six Squadron Mustangs comprising Red, Blue and Green sections proceeded to the Teslin Lake Range each armed with a pair of 500-lb HE bombs. This was the first time HE bombing was carried out during camp. After the ordnance was expended, the aircraft flew to Whitehorse where they were re-armed with the identical bomb load that was then dropped at the Teslin Range on the return trip. Because two camps operated at the same time, there was opportunity to engage in mock bombing or strafing attacks, fighter interception of incoming raiders, and long-range navexes.[28]

As is the case with any undertaking, the deployment to Watson Lake had its light moments. Former Squadron member Maurice Munn learned the hard way that curiosity can prove costly when he asked the wrong question in the wrong place in the wrong time. On a wall in the officers' mess hung a large bell, the purpose of which was not readily apparent. His query as to its function earned him an invitation to try it out. He did. "Well, everybody in the mess came for free drinks and I got stuck with the whole thing. I didn't know there were so many people in there ..."[28] Duff Roblin, who would one day become the Premier of Manitoba, was also in the Squadron at Watson Lake. He brought his bagpipes along, and the reaction of his colleagues was predictable; "there were so many objections to the bagpipes that we finally banished him to the far end of the airfield. On a still night, there was poor old Roblin at the farthest end of the runway twanging away on his bagpipes and everybody very happy that he was hundreds of yards away from the mess and any further damage that he could do to our ears."[30]

BUILDING 200 AT WATSON LAKE; THE THEATRE AND AIRMEN'S MESS, OTHERWISE KNOWN AS "SALOON, DAN MCGREW PROP." YOUNG LAC AL CHALKLEY HANGING OUT

The final sortie of the last day of summer camp at Watson Lake was a costly one for 402 Squadron. F/L John Urquart and W/C Rathwell were conducting dive-bombing at the Teslin Lake Range. The W/C had pulled out of his dive and was orbiting when he noticed an aircraft hit the water, killing the pilot. Urquart's accident was attributed to a wing coming off either in the dive or after pulling out.

F/L JOHN URQUART, WHO LOST HIS LIFE IN A FATAL CRASH IN JULY 1952

During the fourth quarter of 1952, the Squadron busied itself in a wide range of air activities, including simulated attacks on an Army convoy en route to Camp Borden from

Wainwright via Winnipeg, a six-Mustang flypast on Battle of Britain Sunday, cross-country flying, and a heavy concentration on armament practice including cine-camera attacks and two weekends of rocket and machine gun firing over the Shilo range. An extensive instrument flying program was also conducted in September since the Squadron as a whole was in need of additional experience in instrument flying.

In mid-September, 402 was expecting the Calgary squadron to arrive in Winnipeg to participate in the Battle of Britain observances. The Calgarians made every effort to conceal their flight plan in order to keep their time of arrival a secret and deliver a sneak attack. To counter this, the home Squadron sent aircraft up to perform reconnaissance patrols in sections from the U.S. border to half-way up Lake Manitoba and Lake Winnipeg. If a patrol found nothing, it returned to base and other sections took off to take over. The sections flew up and down on daily patrols while the Calgary squadron tried to enter Winnipeg airspace unnoticed. When the intruders were sighted, the patrolling Mustang immediately notified Winnipeg. The defenders then showed up in force and bounced them with their cine gun-cameras in an effort to "shoot" them down. This kind of activity went on all the time among the Squadron pilots themselves, but it was a more enjoyable and realistic exercise to surprise other units.[31]

Another planned attack named "Operation Swift" was delivered on Selkirk later in September by 406 and 418 squadrons' Mitchell aircraft, again escorted by 403 Squadron Mustangs. As in "Exercise Strike," 402 were the defenders under the control of 2402 AC&WU. This operation was immediately followed by a combined flypast over Winnipeg by all 23 aircraft of the participating squadrons.

Yet another combat exercise was in the offing when CJATC Rivers requested that 402 Squadron participate in a joint radar interception and control operation using air-to-air camera gun with the Rivers-based Tactical Fighter Flight (TFF). The target would be Stonewall, Manitoba and 2402 AC&WU would control the interception forces. The TFF would attack from the NW quadrant with four Mustangs in the morning, with 402 Squadron acting as the defending force. In the afternoon the situation would be reversed, hence the operation's code name "Turnabout" which was carried out as planned on 22 November. Also in November, the Langruth air-to-ground firing range that had been established for use by the MacDonald-based No. 1 Air Gunnery School was made available to 402 on weekends. The gunnery targets on the ground counted hits electronically and provided a score – a big improvement over dipping the noses of the bullets in dye. Meanwhile, the Squadron marked its 20th anniversary on 15 November with a good turnout by ex-members and a flying and static display, making it a memorable day for all. Aircraft strength as of this month was ten Mustangs and six Harvards.

Before the year was out, a major improvement was made in the all-important provision of ground personnel to the Squadron. The "Reserve Tradesmen Training Plan (RTTP)

(Basic)" was introduced system-wide in December 1952, superseding the high school student scheme then in place. This move incorporated both students and air cadets, consolidating the majority of aircraft tradesmen training under one plan. The association of air cadets with a Reserve training program would result in the availability of a much greater source of well-trained personnel for the Auxiliary squadrons like 402.

The Reserve Tradesmen Training Plan took precedence over all training at 3052 TTU, since preparation for it in compiling and amending précis, keeping information up to date, and so on accounted for most of the Regular Force time throughout the whole year. Personnel of the RTTP were high school students, some of them serving Air Cadets, 16 to 18 years of age for males, 18 to 21 for females. The annual round went as follows: Recruiting of the young people started in January. Air Cadets applied to the Regular Force Recruiting Centre, and others to the Auxiliary Recruiting Centre at 122 Carlton Street (17 Wing HQ). Phase 1 or Indoctrination Training was carried out on Thursday nights and Sundays until June. Lectures were given on such subjects as dress, manners, the uniform, Air Force law, etc., and instructors from each trade explained what selection of the recruit in his particular trade would mean in the way of training and future employment. Following the July 1 holiday, these new members of the RTTP would report each day at 0800 hours and remain until 1630 hours, Monday through Friday. Before the end of August, 240 hours of trade instruction were to be completed, excluding the time spent on drill, sports and Commanding Officer's Parades. Those who lasted the summer – and most did – would write a Group One trade paper set by Training Command. The purpose of the RTTP was to man Auxiliary units with tradesmen who were up to RCAF standards. It also formed a nucleus of partially trained personnel in case of emergency, and certainly acquainted the youth of Canada with life in the RCAF.

1953

The first half of 1953 saw 402 contribute three Mustangs to "Exercise Assiniboine"; conduct air-to-air firings; share in a flypast, bombing and aerobatics demos ("Operation Air Display") at Regina to mark the expansion of that city's airport; stage a flypast in commemoration of the Queen's Coronation ("Operation Lucky"); and participate in operational exercises "Key Step" and "Key Stone." First on the agenda, however, was "Operation Gunner," a joint armament exercise carried out on 18 January with Mustangs at the Shilo Range at the request of Royal Canadian School of Artillery (RCSA) Shilo. The operation order called for all aircraft to be provided with cine guns and 150 rounds of .5 ammunition per machine gun to two guns on each aircraft; eight 16 ½-lb practice bombs on two aircraft; and eight RPs each for four of the aircraft. The Squadron benefited from the ensuing practice in rocketry, machine gunnery and practice bombing, as well from experience in control by an air contact team. Officers from Tactical Air Group came calling at the home base on 24 April and were pleased with the overall operation and efficiency of the unit. Maintenance was very good during the month and a large number of armament exercises were possible as a result. On 30 April, the radar unit was elevated to squadron status, thereby becoming 2402 AC&WS. The Mustangs conducted a successful bombing control scheme on 7 May in concert with 2402 AC&WS using radar and Identification Friend or Foe (IFF).

A YOUNG PILOT OFFICER UNDERGOING FLIGHT TRAINING DOES SOME EXTRA DUTY LOADING ROUNDS FOR THE MUSTANGS

The Winnipeg Bears went international in mid-June 1953 in "Operation Key Step," a joint USAF/RCAF exercise carried out in the Great Falls, Montana area. A strike force comprising Mustangs from 402 and 403 Squadrons, Harvards from 403 Squadron, and Mitchells from 406 and 418 Squadrons would be pitted against the 29th Air Division USAF, which was responsible for the air defence of the north-central United States. The Americans were equipped with a squadron of F-84Es that could be augmented if need be with F-86D Sabres and F-94s – all jet-powered aircraft. The plan was for the Mitchells to "destroy" the local Anaconda Copper Corporation smoke stack, while the fighters would deliver RP and strafing attacks on the Carter emergency airfield to eliminate all enemy aircraft on the ground or about to take off, and to destroy the administration building and hangar. Following that, they were to act as protective cover for the Mitchells on completion of their bombing run and subsequent landing at the Great Falls airfield. This operation had originally been planned for 6 to 7 June, but serious flooding in and around the Great Falls Air Force Base due to inclement weather necessitated its postponement to 13 to 14 June.

A 402 SQUADRON VAMPIRE IS REFUELLED BEFORE BEING HANGARED FOR THE NIGHT

Meanwhile, Tactical Air Group had been elevated to Command status on 1 June, and the stated task was a familiar one: testing, planning and organizing procedures and techniques that would minimize the problems involved in providing the Canadian Army with air support, particularly in the Canadian North. In wartime, headquarters would control air transport support and tactical air offensive ops. It was originally planned to return to Watson Lake for summer camp, but a polio outbreak in the Whitehorse area necessitated a change in venue. As a consequence, the 5th of July witnessed a contingent of nine 402 Mustangs winging their way eastward. Seven of the aircraft, based at RCAF Station Rockcliffe, would serve as the operational trainers. The other two were flown to Station Trenton to be used as target tugs. Lectures on local flying control and procedure peculiar to the Rockcliffe area were followed by a visit to Station Uplands and inspections of jet aircraft including the Comet, F-86, T-33, CF-100 and Vampire. Then came more lectures, these ones on radio procedure in Air Defence Command (ADC), "pipelines"[32] and how they were operated, ground-controlled interception, high-level battle tactics, the Soviet TU-4 bomber, and various methods of attacking bomber aircraft, including current information on MiG attacks on B-29s in Korea.

Paramount among the lectures received were those on air-to-air firing at the Trenton Range, because armament work was to comprise a major involvement at the 1953 camp. Also on the agenda were ground-controlled intercepts in conjunction with nearby Pinetree Line stations, Ground Controlled Approach (GCA) at Uplands, pipeline approaches and homings at St. Hubert, a practice scramble, cine exercises, local familiarization and formation flying, air tests, and photo interpretation and intelligence. On 10 July, three of the Mustangs were ferried to Station Bagotville for use in "Exercise Tailwind."[33] This was one of a series of continent-wide, joint RCAF/USAF exercises designed to test and train the RCAF's air defence units in as realistic a fashion as possible.[34]

In the usual scheme of things, August was a relatively quiet month for the Squadron as the aircraft underwent inspection and preparations were taken in hand for the upcoming fall training period. However, the Army's Western Command undertook to hold a divisional exercise, code-named "Buffalo IV," in the Wainwright area during the last two weeks of the month and requested air support between the 22nd and 29th. Two 402 Squadron Mustangs were identified for this operation, along with a similar number of Mustangs from 403 Squadron and Mitchells from 418. On a return flight from Vancouver to Winnipeg (a distance of 1100 miles) that same month, F/O Frank Gilland set a Mustang speed record of two hours and forty-nine minutes. Then, a crowd of some 15,000 took in the 19 September Air Force Day activities at Stevenson Field, where the Winnipeg Bears' contribution included a dive-bombing display. It was a busy weekend, for the following day was Battle of Britain Sunday, to which the Squadron contributed a flypast. An extensive armament program was carried out during the month as exceptional weather availed the Squadron of the needed flying time. The program comprised air-to-ground rocketry, gunnery and dive-bombing. Two Squadron aircraft helped kick off the 1953-54 academic year with "Operation Freshie," a flypast and aerobatics display at the University of Manitoba. "In particular," wrote the CO of the University of Manitoba Squadron in his letter of appreciation, "the aerobatics carried out by the two Mustangs from your Squadron highlighted the show. The pilots put on a display to be equalled with the best of flying performances."

A reality check came during "Exercise Lynx-Cat," carried out in the Winnipeg area during the last week of September for the purposes of exercising 406 "City of Saskatoon" (Lynx) Squadron in its role as a tactical strike force, and 402 Squadron in the role of air defence. The objective of the intruders was to sever rail connections between Eastern and Western Canada, and the defending forces were to comprise eight 402 Squadron Mustangs working in conjunction with 2402 AC&WS. The Lynxes were at the top of their game, as no interceptions were made on the six Mitchells that came in low and eluded detection by the Wing radar unit.

Mustang pilots had their share of close calls. In October, F/O Ian Thomson escaped injury after experiencing anoxia at 30,000 feet in a high climb in his Mustang; the aircraft plunged earthward and Thomson came to at 10,000 feet in time to get things under control. His aircraft suffered Category B damage, with the rudder trim and top part of the rudder torn off and the wings stressed and rippled. Hugh Braceland who was the Squadron Chief Engineering Officer at the time describes this incident, or another one like it:

A pilot of a Mustang became hyper-ventilated and found himself in a spiral dive. He managed to pull out but had great difficulty controlling the aircraft. In addition, an oil line had ruptured and he was unable to see much to land. One of the other pilots got into another Mustang and flew about 1-2 feet below the damaged aircraft to see what the damage was – a marvellous feat in itself. Then he led the pilot in for a landing and the aircraft landed safely and taxied in to the ramp. As soon as I looked at the plane I could see there was something terribly wrong. From behind the plane the vertical tail and rudder were leaning to the right by about 30 degrees. On further examination, the whole fuselage had twisted that much.[35]

Bill Davidson recalls

working a weekend and viewing a Mustang in the hangar. Mud jammed in the scoop on the belly, wings with most rivets popped – a real mess. The story I remember was the pilot was over near Beausejour, had oxygen failure at high altitude (this happened more than once), and passed out. He came to in a high-speed dive and pulled out before hitting the ground and over-stressing the aircraft. He force-landed in a field near the town.[36]

W/C Rathwell, DFC, retired in October 1953 as Squadron Commanding Officer, and W/C J.M. Reid, CD, took charge

WINNIPEG BEAR MUSTANG ON FINAL.
NOTE THE LONG-RANGE TANKS UNDER THE WINGS

at the Handing Over Parade on the 8th of the month. Flying was at a minimum during November due to adverse weather. However, that which was accomplished was given over mainly to armament work. One exercise was carried out during the month, "Operation Kangaroo" at Calgary, which called for eight 402 Mustangs to attack and destroy an Army munitions and supply concentration at the Sarcee Range. Four of the aircraft would carry bombs and the rest would provide fighter escort, with 403 Squadron under the control of 2403 AC&WS defending the base in the air. The entire operation gave 402 practice in long-range navigation, bomber escort, locating and bombing a distant target, and unit mobility. The operation upon its conclusion was declared a great success, especially from the maintenance standpoint.

On 16 November, the Squadron again changed commands, moving to Air Defence Command with headquarters at RCAF Station St. Hubert, Quebec. Along with this change, the Squadron was designated 402 "City of Winnipeg" (Fighter) Squadron (Aux). This move was anticipated after the 1953 summer camp, with its emphasis on air-to-air gunnery and instruction in attacks on strategic bombers like the TU-4. Hopes were high that re-equipment with jet fighters was not too far away; in the meantime, Second World War-era piston-driven aircraft were to remain the sole occupants of the Squadron's inventory: at the end of November 1953, the Winnipeg Bears were in possession of 12 Mustangs and six Harvards. Also in November, the Squadron participated in the Remembrance Day Flypast, and four 402 Auxiliary airmen and the OC Support, S/L Barkley, were presented with the Coronation Medal. The four airmen receiving the medals were FSgt Waldie, Sgt Low, Cpl Sawchuck and LAC Fidelak. The year was rounded out in December with "Exercise Cottontail," an Air Force-Army liaison exercise wherein 2402 AC&WS provided communications and VHF contact was maintained at all points of the operation.

W/C J.M. "MATT" REID

"OPERATION FRESHIE"

MUSTANG LINE ON A TYPICAL WINTRY WINNIPEG DAY

For most, "learning" at university meant enrolling in a set number of courses each year and grinding through them until, in the fullness of time, one finally had a degree. For others, the standard academic fare was complemented by military training programs sponsored by the three armed services. The Royal Canadian Air Force version was known as the University Reserve Training Plan. To stimulate interest in the Air Force and to ensure a flow of trained university students as commissioned officers for the Regular service or the Reserves, provision was made in 1948 for the establishment of RCAF (Auxiliary) University Flights at all the major schools across the country. The University of Manitoba Flight was one of the first formed. After a few years, the "Flights" were elevated to "Squadron" status.

The students' training program covered three years. They attended lectures during the academic year, with pay, and could look forward to summer jobs as Flight Cadets while receiving flying or specialist training. For up to 22 weeks during three consecutive summers, they could be employed as pilot, navigator or radio officer trainees, or in eleven non-flying specialist categories: aeronautical engineering, armament, chaplain, construction engineering, education, legal, medical, photography, physical training, signals, and supply.

Each University Flight was to have an establishment of 100 cadets, with selection being made at the rate of thirty-five freshmen annually. At the University of Manitoba (U of M), a Tri-Service Day was instituted as part of Freshie Week. The purpose of Freshie Week was to orient new students to university life; Tri-Service Day served to showcase the military training programs that were available on campus and to provide an opportunity to recruit new cadets. To that end, displays and demonstrations were standard fare, but the 1953 rendition of Tri-Service Day was special.

At that time, the Cold War was up and running and to the locally-based 402 "City of Winnipeg" (Auxiliary) Squadron it was a going concern. The Squadron's role was to repel a Soviet attack across our northern borders; and to deal with such an eventuality, it was equipped with Mustang fighter-bombers. Capable of all kinds of aerial gymnastics, these versatile fighter planes were real crowd-pleasers at air shows, which were staged with great alacrity in those days of concerted military build-up and personnel recruitment. So the idea was put forth: why not give an aerial demonstration over the university on Tri-Service Day? What a splendid opportunity to promote the Air Force in dramatic fashion to a large group of impressionable young prospects for the local university training program.

And so it was that on 23 September 1953, a pair of Mustangs were put through their paces in the skies above the Fort Garry campus. The 402 Squadron Historical Report for the month noted that the event was a "great success"; and afterwards Professor R.C. Bellan, Commanding Officer of the University of Manitoba Squadron, expressed his appreciation to the CO of 402 in a letter which read in part: "In particular, the aerobatics carried out by the two Mustangs from your Squadron highlighted the show. The pilots put on a display to be equalled with the best of flying performances. Their sustained coverage of the campus over the full hour was timely, as the majority of students were able to be out of classes, and see the display."

That is how the 1953-54 academic year at the U of M happened to kick off with its very own air show.

The Air Defence Command Years
(November 1953 - January 1957)

1954

Pursuant to initial postwar plans, the air defence of major cities was the responsibility of Auxiliary flying squadrons, backed up by the mobile Auxiliary AC&W squadrons. This urban defence mandate carried on into the 1950s, even after Regular Force fighter interceptor squadrons had come into being. The evenings and weekends training schedule continued to be standard farc in the Auxiliary, along with the opportunity to take 15 days of training at camp, where they participated in operational exercises with the Regular RCAF and the USAF. A large number of young men were being trained as aircrew each year, on the same courses and with the same privileges as their Regular Force counterparts. On completion of their training, they returned to the Auxiliary squadron that sponsored them, where they were integrated into the squadron's fighting strength. Technical officers could attend specific Regular Force courses in like manner, and airmen had equal opportunities to further their trade knowledge, all receiving full pay and allowances during these training periods. The equivalent of seventy-one days' training did not make a unit fit for all operational duty, but it did narrow the time required by a unit to reach combat efficiency.

On the surface, "combat efficiency" was arguably becoming more theoretical than actual for 402 Squadron in early 1954.

The shortest air routes between bases in the USSR and the industrial centres of southern Canada lay across the polar region, and it was axiomatic that an incoming attack from the Soviet Union should be intercepted as far north as possible. The Winnipeg-based, piston-driven Mustang was not tailor-made for such demands, given the vastness of Canada's North. Even assuming that the Squadron could be deployed to a remote northern airfield such as Churchill and operated from there in time to intercept an incoming threat, the challenges were becoming monumental: the Soviets' swept-wing, long-range Myasishchev Mya-4 jet bomber made its first public appearance over Moscow in May of 1954, and joined the Soviet Air Force DA (Dalnaya Aviatsiya) in 1955/56. This aircraft, with a maximum speed of 620 mph and service ceiling of 40,000 feet, was in an entirely different league from that of the aging Mustang dayfighters. Through 1954 and 1955, Air Defence Command was expanding to a full complement of nine home-based squadrons of long-range, cannon- and missile-armed, all-weather CF-100 Mk 4 jet interceptors,[37] and these, along with USAF interceptor squadrons, were far more capable of dealing with the Soviet bomber threat.

A LINE-UP OF 402 MUSTANGS AND 432 SQUADRON CF-100s

AN AVRO CANADA CF-100 MK V CANUCK: AN AIRCRAFT RUMOURED TO BE RE-EQUIPPING THE SQUADRON

THE WRECKAGE OF MUSTANG 334462 AFTER IT CRASHED AT GLENELLA, MB ON 6 MAY 1954, RESULTING IN THE DEATH OF F/L DELMAR OSBORNE

Be that as it may, the Winnipeg Bears soldiered on with their Mustangs, and interception exercises were carried out with 2402 AC&WS in February of 1954. Unfortunately, the first half of that year proved to be particularly costly for the Squadron: F/O J.A. MacLennan died in a crash near Teulon while on a cross-country flight; F/L D.L. Osborne lost his life while returning to Winnipeg on the last leg of a cross-country to Rivers and Dauphin; and F/O Ian Thomson was seriously injured when he crashed while carrying out GCAs. As an element of Air Defence Command, there existed the prospect of 402 receiving jet fighters in due course to replace the obsolete Mustangs. To that end, all of the Squadron pilots underwent decompression chamber runs and received lectures on the physiological aspects of high-altitude flight at Station Portage in early June in preparation for conversion to jet aircraft. Back on home turf, 402's rocket-firing display on Air Force Day went over well before an appreciative multitude, as was to be expected. Fortunately, F/O Toews' Mustang received only minor damage when the Army demolition specialists blew up the target just as the aircraft passed over it in the final attack. Tom Patterson describes the incident as follows:

We were firing rockets on Air Force Day here in Winnipeg. Someone thought it would be a unique demonstration by having the Army build a small wooden shack of 2x4 lumber covered with cardboard, with a detonator and small explosive charge placed inside. We made several passes with our Mustangs and fired rockets having inert warheads. The aircraft directly behind me was my number two, Art Toews. On the final pass he fired his rockets prematurely and they passed underneath my aircraft. So I fired my rockets. As I pulled up, Art followed behind me. The Army had placed a larger charge than anticipated, and when they blew up the shack, something hit Art's spinner with a thud and ricocheted off. After he landed his mud-splattered Mustang, it took him a while to climb out of the cockpit. A piece of wood had hurtled up from the explosion and hit his spinner, leaving it dented.[38]

A 402 MUSTANG AT THE UNHAPPY END OF A WHEELS-UP LANDING

Eight Mustangs were flown to RCAF Station St. Hubert on 3 July for the 1954 summer camp, while the balance of the aircrew and the groundcrew were ferried down by Dakotas of the Winnipeg-based 111 Search and Rescue Flight and C-119s from Edmonton. The first few days were given over to ground-controlled intercepts and pipelines preparatory to the big show, "Exercise Checkpoint," carried out in conjunction with the United States Air Force (USAF). This was in fact the third joint RCAF/USAF exercise to be staged in recent years, although the first in the series to actually involve 402 Squadron operating alongside the USAF.[39] The Canadian portion was designed to exercise Air Defence Command. In all, five Regular (four CF-100, one F-86) and eight Auxiliary (four Vampire, four Mustang) fighter squadrons participated, while personnel from eleven Auxiliary AC&W squadrons augmented the Regular radar force. The intelligence organization was entirely manned by members of the Auxiliary. 402 Squadron pilots made several kills, including a B-36 and a civilian seaplane, which, though not claimed by any pilot, showed up when the cine film was assessed.

Overall, "Exercise Checkpoint" saw significant improvements over its predecessors. The quality of the Regular and Auxiliary aircraft controllers was showing marked improvement; one Sector Commander was particularly pleased with the Auxiliary controllers, who had completed 86 interceptions compared to 14 for the Regular Force in his sector. The Auxiliary were lauded for their efforts despite the fact that they had been given no warning on the nature of the exercise and no time to prepare. Thus, as time went on and the threat from the Soviet Union was perceived to be increasing, the performance of the Auxiliary as a whole was also improving.[40] While at summer camp, W/C Reid and S/Ls Gray, Patterson and McMillan flew to Station Trenton to participate in the MacBrien Trophy Shoot, an annual air-to-air and air-to-ground marksmanship competition among the Auxiliaries.[41]

RESPLENDENT IN FULL 402 WAR PAINT, AND POLISHED UNTIL THE ALUMINIUM SHONE LIKE SILVER, THIS T-33 HAS ONE OF THE MOST ATTRACTIVE SCHEMES EVER APPLIED TO THE FAMOUS JET TRAINER.

The move toward jet conversion in 402 Squadron proceeded apace. In early September, F/O F. Russell departed for Station Chatham for a course on T-33s in anticipation of the delivery of two such aircraft to Winnipeg. The Squadron received the T-33 conversion syllabus of training "with great delight,"[42] and ground training began with lectures in airmanship, engines, and airframe and engine handling. By the end of September, three of the Regular Support officers had been converted to jet aircraft, as had most of the younger Auxiliary pilots, either at Gimli, Portage or MacDonald. During two weekends in November, eleven Auxiliary pilots and one Support pilot attended the Fighter Technical Training course on T-33 aircraft at Gimli. This course was considered a "must" for those about to undergo jet conversion, and the 402 pilots found it to be highly beneficial. The big day came on 19 November, when the first T-33 was flown by 402 Squadron. The Public Relations Officer and three members of the press were in attendance for the occasion.

WINTER START ON ONE OF 402 SQUADRON'S TWO T-33s

However, the fourth quarter of 1954 was not all about conversion to jets. Public appearances included a flypast over the Legislative Building during Battle of Britain Sunday services, and a small, three-plane show over the University during the annual recruiting drive. On the operational side, five Mustangs, followed by two Dakotas from 111 SAR Flight carrying extra aircrew and ground personnel, proceeded to the Pilot Weapons School for a day for "Operation MacDonald." In this exercise, five Mustangs were detailed to take off for the Langruth Range with 200 rounds of .5 calibre ammunition, carry out air-to-ground firing, and return to Station MacDonald for rearming with machine gun ammunition and rockets. Twenty-two sorties in air-to-ground machine gun firing, and eight on rockets, were carried out. The results were deemed to be very good, with an average success rate of about 25%. One Harvard was ferried to Station MacDonald to give flights to groundcrew when off duty during the day. The exercise ended with S/L Gray leading a section of three aircraft to Virden, Manitoba for a formation flypast for the opening ceremonies of the town's airstrip.

During the first week in November, seven Mustangs led by W/C Reid, proceeded to Wold Chamberlain Air Force Base, Minneapolis, on "Exercise Cross Border." As background, a state of war had existed since 30 October 1954 between "Northland" and "Southland." Northland undertook to launch an attack against Mid-West Southland with the limited aim of penetrating the defences of that area's major population centre, Minneapolis. En route to and from the American city, the USAF carried out ground-controlled intercepts on the 402 aircraft. In addition to being of great training value, exercises of this kind were a significant morale-builder for the airmen, as shown time and again by the numbers of volunteers that consistently exceeded operational requirements.

WINNIPEGGER BOB MAY'S MUSTANG, BEAUTIFULLY RESTORED IN THE "CITY OF WINNIPEG" SQUADRON'S COLOURS

1955

The year 1955 started off with something special – the Squadron Pipe Band became active on 15 February. By the end of June, it had won a competition against other pipe bands in Edmonton, with individual members receiving additional prizes as well. A further measure of the quality of the band and the enthusiasm of its members can be gauged from the fact that in May of the following year, it won the Western Canadian Pipe Band Competition in Saskatoon. Of first-class calibre, the band was a considerable credit to the RCAF. Operationally, the Squadron's sustained commitment to tactical air support manifested itself in the two-day "Operation Centre Punch," the first phase of which took place on 19 February 1955. An armoured and an infantry regiment were ensconced in South and East Shilo Camp, and it was the role of 402 Squadron, with RCAF Station Winnipeg as its base of operations, to attack and destroy enemy forces with machine guns and rockets. Pre-flight briefing and air control and radio communication at the range were the responsibility of the Army's No.2 Air Liaison Section. In the event, weather forced postponement of the second day of the program, but this was made good in April with six Squadron aircraft operating out of Rivers. During this phase, the Mustangs fired rockets, dropped napalm and dive-bombed with 500-lb HEs before returning to their Winnipeg home base.

A LINE-UP OF 402 AIRCRAFT – AT LEAST ONE HARVARD AND FIVE MUSTANGS

The Squadron's two T-33s were showcased in a ground display at their hangar as part of the annual Air Force Day program in June, while six Mustangs participated in a formation flypast for the event. Armament work was high on the agenda at the 1955 summer camp, which took place in July at RCAF Station Cold Lake, Alberta, home of No.3 All-Weather (Fighter) Operational Training Unit with its CF-100s, the type to which the Winnipeg Bears hoped to convert in due course. In November, a flight of four Mustangs provided a flypast to the dedication of Andrew Mynarski, VC, School, named in honour of the Winnipeg war hero. It was then off to Edmonton in early December, where more than two hundred officers and men of the Winnipeg, Saskatoon, Calgary, and Edmonton Auxiliary squadrons participated in a weekend-long training exercise. The Squadron's seven Mustangs and two T-33s, along with their 403 Squadron cohorts, busied themselves flying escort and interception for the Mitchells from Edmonton and Saskatoon.

1956

By the mid-1950s, the direction of the Cold War was changing. It was becoming increasingly unlikely to military planners that enemy land forces would attempt to gain a foothold on Canadian soil. Rather, manned bombers and missiles were perceived as the over-riding threat, in consequence of which priorities were shifting from an Auxiliary-dominated line of defence to a substantial Regular Force of full-time professionals. "Changing technology had made it probable that war, if it came, would be fought with immediately available resources, and would be over before reserves could play their traditional part."[43] Furthermore, the Army was becoming less interested in tactical offensive air support and more favourably inclined toward air transport. In sum, changes were in the offing for the citizen air force, and for 402 Squadron the changes came sooner than for most.

1956 did indeed prove to be a watershed year for the Winnipeg Bears. First on the agenda was the introduction on 5 January of a "pre-aircrew" training scheme intended to maintain a more ensured flow of Short Short Service Commission (SSSC) pilots into the Squadron. Drawing heavily on RTTP personnel already on the Squadron or other Wing units, the yearly program consisted of two 3-month courses. Twenty-four hours of ground lectures and twelve hours of air experience per course were deemed sufficient to bring candidates up to wings standard. A second reason of equal importance was to enable the Squadron to better determine the candidates' motivation and suitability for aircrew training. In the past, it had been all but impossible to select the best-suited personnel for training due to the fact that a candidate was interviewed for one hour and then launched into his aircrew training. Under this scheme, the CO was unable to accurately assess the candidate's potential, resulting in an extremely high wastage rate.

THE "SARGE" POINTS OUT THE SNAG TO THE "FLIGHT"

CHAPTER 3

AN EXPEDITOR OVER THE PRAIRIES

But by far the most radical change in the Squadron's fortunes came with its relinquishing of the fighter role and its replacement with one of air navigator training. The hoped-for shift to CF-100 jet interceptors failed to materialize: not only had the Regular Force been sufficiently built up to handle the demands of air defence, but maintenance of the CF-100's electronics system was considered beyond the capabilities of the Auxiliary units. Accordingly, plans were made to convert from the T-33 and the Mustang to the twin-engine C-45 Expeditor, the RCAF's première twin-engine navigation trainer. The news was greeted with mixed emotions: gone were the glamour days when the Squadron functioned as a combat unit. June 1956 was the last month 402 operated in a fighter squadron capacity, although it remained for a time as an element of Air Defence Command. The transition to the C-45 was actually made on somebody else's aircraft: the Winnipeg Bears used No.2 Air Observer School machines for both ground and flying training at the outset, still in 2 AOS markings but with the City of Winnipeg crest on the entrance door.[44] When 402 got its own aircraft, the "AC" identification letters were applied to them. The Squadron diary for June 1956 notes that as of the 20th of the month it was now designated "No.402 Navigation Training Sqn (Auxiliary)", and with it came an expectation that the number of Regular Force groundcrew would be substantially reduced.

The 9th of June was Air Force Day in Winnipeg. The Squadron's contribution was six Mustangs, led by W/C Gray, in a series of flypasts in formation, the last one ending in a modified version of the Prince of Wales Feather, which seemed to greatly impress the adoring public. RCAF Station Saskatoon was the venue for the 1956 summer camp, the prime purpose of which was to provide Squadron personnel with ground and air conversion training on the Expeditor. The use of crews as such was a new experience, as the Mustangs and the Vampires before them were single-seaters. Training of pilots, co-pilots and navigators would now be the order of the day. On 19 July, the T-33s were transferred to Toronto. The Mustangs were retired in September with the majority, (7-9 in number) being tied down at the north end of 1 Hanger with the rest (about 3) being sent into storage at the Carberry aerodrome.[45]

"HAVE I GOT A DEAL FOR YOU" RETIRED 402 MUSTANGS FOR SALE.

By the end of August, the Squadron had taken delivery of eight Expeditors and the pilots applied themselves to earning their instrument ratings. In the meantime, 402 had undergone another change of command: W/C Reid, CD, had retired effective 14 May 1956 as Squadron CO, and was replaced the following day by W/C D.M. Gray, CD. It would appear that the role of the Squadron was up in the air for a bit, notwithstanding the notation in the 20 June diary that it was now a navigation training unit – a 6 September 1956 entry in the Historical Record reads: "The official role of the Sqn has not yet been officially stated, although it is expected to be either light transport or the training of navigators."

Since the time of his adoption as the Squadron mascot, Sergeant Minto had been patiently carrying out his daily duties of meeting the public at the Assiniboine Park Zoo. By the time 1956 had arrived, the College of Arms had decided that his face on the Squadron crest too much resembled that of a wolf. On this account, a replacement emblem was deemed in order, and the crest finally approved by Her Majesty and registered with the College in November 1956 displayed a totem bear figure in the artistic style of the Aboriginal peoples of the West Coast.

THE UNOFFICIAL 402 SQUADRON BADGE, CIRCA EARLY 1950s

Chapter 3

The Training Command Years
(January 1957 - March 1961)

1957

With its new role and equipment, 402 Squadron was noticeably out of place in Air Defence Command, and, on 29 January 1957, the Squadron was transferred to Air Training Command and re-designated "402 'City of Winnipeg' (T) Squadron (Aux)" (although 2402 AC&WS and 4003 Intelligence Unit would stay put). The "T", however, stood for "transport", not "training"; 402 was, for now, a transport squadron within a training command. The first year within the new scheme of things would be taken up exclusively with transport training on the Squadron, and the results were not long in coming. Between 25 and 28 April 1957, the Multi-Engine Flight of the Central Flying School tested all 402 pilots, Auxiliary and Regular, on their flying proficiency and knowledge of the Expeditor. Also checked were such related subjects as flight safety, compass swings, navigation training, and flying equipment. The results were highly satisfactory to all concerned; especially considering the unit's limited experience on the type, and the recommendation was that the Squadron pilots' overall proficiency demonstrated a capability for limited transport duties at the discretion of the Commanding Officer.

In March, the Winnipeg Bears said good-bye to an old friend as the Harvard was retired from Squadron service. Having been part of the UE for almost eleven years, the Harvard would prove to be the longest-lived aircraft type on the 402 inventory while the Squadron was a component of the RCAF. Most were Mk IIs; the exceptions were a pair of Mk 4s that were brought on strength after that variant had entered service with the Air Force in late 1951. The record shows that no fewer than 23 Harvards served with 402 Squadron during the unit's postwar RCAF days.[46]

402 MK II HARVARD IN THE EARLY SCHEME

In view of the changing concept of the role for Auxiliary flying formations in the RCAF, the 1957 summer camp was particularly significant for the Winnipeg Bears. The Squadron and the Regular Support Unit proceeded on 29 June to Station Gimli for training until the 13th of July, during which time the aircrew concentrated on building up experience in long-range flying under Instrument Flight Rules in transport-type aircraft. For the first five days, seven Expeditors were on strength. For the remainder of the camp, eight aircraft were available. The majority of the trips were made under IFR flight plans, and aircraft landed at Armstrong, Calgary, Edmonton, Kapuskasing, the Lakehead, London (Ontario), MacDonald, Montreal, North Bay, Ottawa, Saskatoon, St. Hubert, Toronto, Trenton, and Val D'Or. Excellent experience was obtained in various types of weather conditions ranging from VFR to IFR minima. One navigator was in attendance, and four long-range flights specifically for navigator experience were carried out. The groundcrew delivered the aircraft as required (the serviceability rate was 91.8%) and aircrew maintained a sustained effort to utilize them to the fullest. Training for maintenance personnel consisted of one week of concentrated ground school under the direction of 3052 TTU, and one week of practical experience. To further utilize the Squadron's capabilities, the transportation requirements of 2402 AC&W Squadron to RCAF Station Senneterre were undertaken. The flying experience gained demonstrated that the Squadron had the necessary potential and capability to function in an effective transport role. In further keeping with the latter, and with an eye to background knowledge for future operations, long-range training flights were carried out in September to Regular Force transport squadrons based in Edmonton, Trenton, Montreal and Ottawa to check on the organization and operations of those units.

SGT J. JANSKOSKI PERFORMING A THOROUGH CHECK ON THE UNDERCARRIAGE OF A SQUADRON EXPEDITOR AT GIMLI

The new face of Army co-op was in evidence with "Exercise Goldeye", conducted in late September. The scenario was as follows: A recent enemy air strike had caused considerable damage to bridges and approaches in Manitoba. Friendly forces had subsequently established air superiority, and it was possible for transport aircraft to operate in the area from Flin Flon to Winnipeg. The Squadron's mission

ON THE RAMP IN WINNIPEG

was to transport 18 reinforcement Royal Canadian Engineers from Flin Flon to Winnipeg, and to return them to Flin Flon once their tasks were completed. Six Expeditors were laid on for each trip, and the exercise was carried out as planned without incident.

The Squadron's training program in navigation unofficially began at the beginning of November, although this work in the air was held up during the month due to weather. The year 1957 ended on a sad note with the death of Flying Cadet K.H. Irvine, who was killed in a flying accident at RCAF Station Penhold. F/C Irvine had been sponsored by the Squadron and was undergoing pilot training at the time.

1958

With the year 1958 came a complete change in the mandate of the RCAF Auxiliary. The government announced that the majority of the squadrons would convert to an emergency and rescue role, the prime concern now being that the Cold War would become hot and a nuclear holocaust would ensue. The government and armed forces had already begun preparations for a military response in aid of civil authorities in the event of a nuclear war with the resultant civilian casualties. The Auxiliary, which had been considered a pool of trained air and ground personnel ready to be absorbed, if necessary, into Regular Force

EXPEDITOR PILOTS – L TO R: REID SMITH, IAN PATRICK, GLEN DOWNES, ARCHIE GITTAL, FRED MILLER

fighting units, now would be an entity unto itself. Its role in general terms read as follows: "Auxiliary squadrons are established to meet short-haul military air transport requirements, air transport for reserve units and air cadets, and assistance in search and rescue activities."[47] For the Winnipeg Bears, as of 1 April 1958, the new policy translated into air operations in support of military and civilian requirements. As an emergency and rescue squadron, 402 would co-operate in peacetime with the Civil Authority and provide aerial observation to determine the condition and extent of disaster areas. The previous role of transportation of medical personnel and supplies, evacuation, and search activities would be carried on as usual, and peace- and war-time operations with the military (i.e., short-haul transportation and search duties) would carry on as before.

The most notable role change concerned wartime operations with the Metropolitan Civil Defence Organization. This would include giving authorities confirmation of ground zero and assessments of damage in the bomb area, aerial radiological monitoring within the fallout pattern, reporting the extent of the fire zone, evacuation route reconnaissance, provision of transportation for relief columns, and communications. Since 402 Squadron had already relinquished its combat orientation, the transition to the national survival set-up as an Emergency and Rescue formation was not quite as abrupt for the Winnipeg Bears as it was for the fighter and light bomber squadrons, nonetheless, a new training regime had to be put in place to these ends. However, with mobilization no longer a prospect, the eventual demise of 5001 Intelligence Unit was only a matter of time.

EXPEDITOR TO MINNEAPOLIS

In that same month of April 1958, the Multi-Engine Flight of CFS Trenton visited 402 for the purpose of checking the proficiency of Squadron pilots and observers. Eighteen check rides were carried out, with satisfactory results. A planned-monthly Squadron newspaper, *Totem Talk*, came into being at the end of April, although a cantankerous printing machine lent truth to the maxim "good things don't come easy."

In early May, the Bears were called upon to transport eight Army personnel and their baggage on a round trip from Flin Flon to Brandon as part of a combined civil defence exercise. On the last day of May, the Squadron CO attended an Auxiliary Commanding Officers' conference at Air Force Headquarters, where he learned that a plan had been formulated to complement 402's Expeditors with de Havilland Canada Otters. The UE would comprise four aircraft of each type. Along with new aircraft would come a new plan for the way in which they looked: In July of 1958, the last vestiges of the old International Civil Aviation Organization (ICAO) system were abolished. For the Squadron, this meant the disappearance of the distinctive "AC" unit identifier and its replacement with the letters "RCAF." Once the new marking system was in place, it would no longer be possible for the casual observer to determine the aircraft's unit simply by looking at it.

With late June/early July came the annual summer camp and two weeks of intensive air and groundcrew training, again at *Gimli*, with a flying schedule focused on Captains Proficiency Checks, Long-Range Training Flights, and Category (Instructor) and Instrument Checks. The results of this training indicated that the Squadron was capable of maintaining a light transport role on a sustained basis should emergency conditions require it. Accompanying the Squadron this year was 4003 Medical Unit for contact training duties. The CO declared the performance of Auxiliary groundcrews during this camp to have been most satisfying; approximately 90% of the total workload was carried out by Auxiliary tradesmen. Regular Support personnel did the remainder where Auxiliary strength was numerically weak. It was considered that if this deficiency could be overcome in time, there would be no reason why Regular Support airmen could not remain at the Squadron home base when the unit moved to the next summer camp.

On 6 and 7 December, the Squadron once again airlifted Army personnel to the North on a civil defence exercise ("Operation Goldeye II"), this time to The Pas and back.

1959

It wasn't quite Hollywood, but in late April 1959, a pair of officers directed the production of a film depicting typical Squadron life. The plan was to feature it as part of the static display on Air Force Day scheduled for 8 August.

W/C GRAY AND S/L MCMILLAN CONDUCT A BRIEFING, CIRCA 1959

TRANSFERRING CASUALTIES TO A WAITING AMBULANCE

On 4 July, eight Squadron Expeditors and two Dakota aircraft departed Winnipeg for *Station St. Hubert* and summer camp. The mission of this year's camp was to provide a period of intensive aircrew and groundcrew training appropriate to the role of multi-purpose and light transport. This translated into long-range cross-country pilot and navigator training, practice searches, visits to Air Defence Command formations and to an aircraft manufacturing plant, and lectures and films on nuclear warfare. Number 4003 Medical Unit would be associated with Station St. Hubert Medical Section in carrying out contact training and rendering other such services as may be required. Regular Force medical staff provided practical training in casualty evacuation and aero-medical subjects. An intensive groundcrew training program was carried out with emphasis on the practical phase, and two aircraft were allotted solely for this purpose.

To acknowledge the support received by the members on the home front, a "Family Day" committee had been struck, and on 25 October, the first Squadron Family Day was held. The families of the members were invited to come and visit the hangar, where documentary films and cartoons provided the entertainment and toy gliders and balloons were handed out to the children. A helicopter from 111 KU staged a simulated rescue, and F/O Henderson, also a member of the Winnipeg Flying Club, provided flights for the youngsters in one of the Club's aircraft. The success of the event was such that plans were made to have it become an annual feature.

1960

Because of the ever-increasing amount of civilian light-aircraft flying, 402 was frequently being called upon to assist in air searches. With this demand came the requisite training, and on 21 February 1960, "Operation John" was put into effect. The purpose of this exercise was to simulate a search for a downed aircraft. It began at 2105 hrs the previous night at the Officers' Mess at 17 Wing Headquarters, where most of the crews were assigned; the remainder were contacted at home. Briefing was held at 0630 hrs, and the first aircraft was in the search area by 0800 hrs. The "lost" aircraft, a derelict Lancaster, was sighted and the smoothly and efficiently functioning operation was declared a success, with similar searches being proposed for the future.

EXPEDITOR PILOTS F/L E.R. WOLKOWSKI AND S/L W.C. PACHOLKA (LEFT)

March 12th 1960 marked the beginning of the Nuclear Defence Courses for RCAF (Auxiliary) personnel at Mawdesley Hall, Air Observer School, RCAF Station Winnipeg. The purposes of the course were to familiarize 17 Wing personnel with the effects and peculiarities of an atomic blast and the measures to be taken in the event of such a catastrophe. From 7 to 14 April, yet another staff evaluation party visited the Squadron, this one from the Air Transport Command Assessment Unit. Air checks and ground examinations were given to available aircrew, on the basis of which the Squadron was declared "operationally acceptable."[48] In conjunction with the new civil defence role, the Squadron at long last received its first de Havilland Otter aircraft on 12 May, and training programs on type were quickly set up for flying and technical personnel. A more versatile machine that was better suited to the needs of the Squadron; the Otter greatly increased the scope of the Winnipeg Bears' operations. Before the month of May was out, 402 Squadron aircraft took part in "SAR Harrison": An aircraft had gone down in the vicinity of Flin Flon, and the search aircraft flew out of The Pas to their designated search areas. Summer camp in 1960 was held at Calgary, where the Squadron underwent dress rehearsals in its new role.

THE INTERIOR OF AN OTTER LOOKING FORWARD INTO THE COCKPIT

A 402 SQUADRON OTTER –
NOTE THE WINNIPEG CENTENNIAL CREST ON THE PILOT'S DOOR

Former CO Ernie Harris provides a first-hand account of the Expeditor/Otter days in the Squadron:

> When I first started, we were flying the C-45, the Expeditor – some people call it the Beech 18 – and we used to have service flights between Winnipeg, Saskatoon and Moose Jaw. We carried either passengers or freight – whatever they wanted. We were also on search and rescue; we did the Santa Claus drop to some of the reserves in the North at Christmas time; and of course we could be tasked to take passengers or freight anywhere in Canada, very seldom into the States. But anywhere in Canada; flew the Expeditor over the Rockies. Flew it as far west as the Rockies and as far north as Inuvik in the Northwest Territories for what we called the "summer concentration" or the summer camps, and at that time we had both the Expeditor and the single-engine Otter. We were up there with both aircraft to try and find out where any of the landing strips were and to update any information on some of them that the Department of National Defence knew about because as you know, especially up in the North, sometimes there is an airstrip and two or three years later there isn't one: things have grown in, or maybe two or three years later is improved – they put lights in, etc. So we were updating this information and of course, to make a bigger appearance of military in the North; and you hear it even today, sovereignty far up North, and even further north.
>
> We had one Otter on floats and the rest on wheels during the summer time, and these Otters could go into very short fields. The Expeditors of course couldn't go into quite as short fields and had to have landing strips that were a little better conditioned than the Otters. The Otters could land in almost any farmer's field really – probably stretching that a bit – but it could do that and the Expeditor was a little faster, probably about 150 knots average: wind would affect it, increase it or decrease it. My preference was probably for the Expeditor because I have an awful lot more hours on it. But the Otter was a real, real challenge because it flew quite a bit differently when it was empty as to when it was loaded. It was almost like learning how to fly a new aircraft! Because of its being slow, that was the only handicap. But because it was quite slow, it could go into places that a lot of other aircraft couldn't – the Expeditor couldn't go in because of its undercarriage construction – and so it could receive much more punishment from the landing fields than the Expeditor.[49]

On 17 July 1960, W/C Gray, CD, retired as CO of 402 Squadron and transferred to 17 Wing Headquarters. The following day, S/L J. T. "Tom" Patterson, CD, assumed command with simultaneous promotion to the rank of Wing Commander. A handing-over parade was held in No.1 Hangar on 13 October.

1961

For some time now, the Squadron had been tasked with the responsibility of carrying out monthly flights to RCAF Station Winisk, on the shore of Hudson Bay, to provide that outpost with the services of a clergyman, a task hitherto filled by 111 KU Flight. It was during one of these trips that the Squadron suffered its first Expeditor crash. On 7 January 1961, while en route to Winisk via Armstrong, Ontario, the aircraft lost both engines at an altitude of 500 feet on final approach to Armstrong and was written off. Fortunately, all five people on board, including a chaplain, escaped fatal injury.

On 21 January, Squadron CO W/C Patterson outlined an organizational plan for "calling out to service." Nicknamed the "402 Scramble Plan," it was devised to deal with emergencies wherein personnel are called upon to perform Air Force duties (e.g., air search, forest fire surveillance, etc) as may be directed by a designated military authority. It involved a three-phase action sequence as follows: (1) on receipt of a warning or alert, the Regular Support Unit (RSU) officers would notify key Auxiliary personnel; (2) they in turn would notify Flight or Section commanders, who (3) would contact members of their sections, as required, ensuring an effective contribution to their assigned function.

SERGEANT W. FIRTH; SERVICING WITH A SMILE AND A WAVE

Marshalling the needed manpower for Squadron operations always had its share of challenges as Ernie Harris recalls:

> The flight plans for some of the senior officers probably couldn't fit their schedules, you see, so they would task us to do it, and naturally we would if we could get people to fly. Weekends were no problem; we had all sorts of people who could fly on weekends. But if it was during the week …we did have two Regular Force pilots most of the time, but they had a terrific amount of administration work to do also, and of course they had to have their days off too. With the Expeditor aircraft, the rules and regulations were that you needed two pilots, so if you put both of the Regular Force pilots on for a trip that would take them away for a few days, well then it left you short during that week.

Nobody likes to work any longer than your eight hours and five days a week, so if you have your Regular Force coming out on the evening that you're working, you know they're there for four or five hours, and of course if they've been flying all day, etc., they must have their time off. There's no way you can pay any overtime in the Services; the best you can do is try to give them one hour off for every hour they work overtime. In civilian life you know that if you work overtime, you want time-and-a-half or double-time. You can't do that in the Services, so there are people that didn't like to work overtime, but it was necessary sometimes.[50]

As of March 1961, the eleven existing RCAF Auxiliary squadrons were variously distributed among two separate Commands and one Division.[51] On 1 April, they were all brought under the control of Air Transport Command (ATC), a change especially appropriate to their current role. In the meantime, the advent of the Semi Automatic Ground Environment (SAGE) system in the Regular AC&W Pinetree Line squadrons meant that their back up Auxiliary counterparts were no longer required, and Winnipeg's 2402 AC&WS was disbanded on 31 March 1961.[52]

The Air Transport Command Years (April 1961 - February 1968)

By June 1961, 402 Squadron had been relocated to No.4 Hanger and in the early summer of that year, several of the Regular Support maintenance crews were tasked to make the remaining Mustangs airworthy sufficient to be flown out by contract pilots. According to the memory of Ron Salome, "I remember starting them up, and having them taxi away. One pilot brought the aircraft back, shut it down, and said it scared the heck out of him. When he put the power to it, the P-Torque was too much for him to handle. What he didn't know is as soon as you got if off the ground it was a pussy cat."

RUNNING UP MUSTANG 297'S MERLIN

Summer Camp 1961 was held at Winnipeg from 1 to 15 July, in an effort to make it more convenient for Squadron personnel to get away from their civilian jobs. In addition to 402 and 17 Wing HQ., the camp was supplemented by elements of 4003 MU and 3052 TTU. Comprehensive programs were carried out involving the respective trades of aircrew (instrument rating training, Captains' route checks, pilot proficiency checks, low-level and precision long-range navigation, night flying, simulator training, lectures on nuclear warfare) and, groundcrew (on-the-job training). The aircrew component covered the phases of practical training such as transport operations, conversion and search exercises. The story of the last-mentioned is particularly interesting: "SAR Totem" was a simulated search for a grounded Otter aircraft in the local area. This exercise was temporarily abandoned when an actual search, "SAR MacLeod," was carried out at the request of 111 KU. The missing aircraft was located in a lake where it had gone down. A most notable feature of "SAR MacLeod" was the rapidity with which a sighting was made by a 402 aircraft; only 7.57 hours had elapsed from the moment that 402 was ordered into the search until the lost aircraft was sighted. Upon completion of this effort, the simulated search was resumed and carried to a successful conclusion. The Operation Order for Summer Camp 1961 also called for assignment of two Otters and supporting personnel to Proctor Field at Camp Shilo for the periods 3 to 6 July and 10 to 14 July, to participate in area reconnaissance, casualty evacuation, Natural Survival exercises and provide transport for 19 Militia Group and the School of Artillery.

November 25th 1961 proved to be one of the big dates in the history of 402 Squadron. With colourful and traditional ceremony, the Squadron received its own Standard from the hands of the Lieutenant Governor of Manitoba, the Honourable Errick Willis, at No.10 Hangar, RCAF Station Winnipeg. The Queen granted the honour of possessing a squadron standard to those units of the RCAF that had been existence for 25 years or had distinguished themselves by outstanding operations. The design of 402's standard included the Squadron's Totem Bear badge surrounded by scrolls bearing its battle honours of the Second World War.

THE 402 SQUADRON COLOURS

THE CONSECRATION OF THE COLOURS 25 NOVEMBER 1961. WO1 FRANK WALDIE IS HOLDING THE STANDARD WHILE WO2 WILLIAM VAREY ASSISTS

1962

The year 1962 was not a month old when the Winnipeg Bears were involved in yet another search. Shortly after reaching their assigned area in "SAR Dobbs," the First Officer of Expeditor 2328 spotted one of the occupants of the downed aircraft. The individual was observed on a frozen lake some 120 miles northwest of Winnipeg. The Expeditor circled the man until an Otter from 111 KU reached the area and let down on the lake. The Expeditor then proceeded to follow the snowshoe tracks of the sighted survivor and located the downed aircraft about 15 miles due east of where it had ground-looped during an unscheduled landing.

On May 26, four Squadron Expeditors and an Otter provided a flypast display for the official opening of the new Flin Flon Airport. After the flypast, one Expeditor and the Otter landed to provide a static display for the public. It was back to summer camp once again from 30 June to 14 July, this time at CJATC Rivers. The camp was attended by 17 Wing HQ, 402 Squadron and 4003 MU, and it may well have been this occasion that Al Henderson was speaking of when he recalled:

> In 1962 the Squadron went out to Shilo with two [Otter] aircraft and we worked with the militia – different exercises. We did our operating out of the Rivers air base and we'd fly over to Shilo in the morning. Basically we were operating airlift. We would take troops into various spots throughout the Camp Shilo area and drop them off and let them do their procedures and exercises. We also did some exercises with our own medical group in Shilo: the set-up was that there was to be an attack on a station, and a medical unit would go in and remove the injured and we would fly the Otter which was set up as a stretcher carrier and airlift these people out of various spots throughout Shilo,[53]

W/C Patterson was transferred to 17 Wing and was succeeded as Squadron CO by W/C D.R. Scott, CD, effective 16 October 1962. In November, a most interesting presentation was made to 402: Mrs. C.E. O'Grady, the last President of the RCAF Women's Auxiliary, turned over a pair of Dutch work shoes ("clogs") that had been sent to her organization from Holland in 1945 in appreciation of the wartime support it had given the Squadron. The shoes, autographed by Squadron members before being sent to Canada, formed a historic and unique record of 402's wartime membership and remain one of the Winnipeg Bears' prize possessions.

OTTER PILOTS, W/C GRAY IS ON THE FAR RIGHT

1963

The change-over to Air Transport Command two years previously had brought with it added duties as time went on. In January of 1963, 402 assumed yet another monthly commitment: a "Flying Doctor Medical Service" under which Squadron and 4003 Medical Unit personnel were transported to Pinetree Line stations Armstrong and Gypsumville, and any other establishment that required medical assistance. This task was additional to the regular duties of emergency air evacuation and search-and-rescue operations, not to mention training in basic survival and ground search, radiac[54], first aid, area medevac, and summer camp Army co-op at RCAF Station Gimli, all of which occupied Squadron members and their six Expeditors and two Otters during 1963.

From Ernie Harris comes an excellent commentary on the summer camps and a tribute to the Squadron's "people behind the scenes":

> Summer camp was a great morale-builder for the groundcrew. As everybody knows, the aircrews, especially the pilots, get all the glory and such as that; but really, for any squadron to operate, the man or the woman who sweeps the hangar floor is just as important, if maybe not more important, than the man or woman who flies the aircraft. But they don't get the glory, these people who sweep the floor, who service the airplanes, the girls or the typists, the clerks, because it's always been the person flying the aircraft that gets the glory. And really it's the people who support – if they didn't ever sweep the hangar floor, you're never going to have aircraft flying. So they are just as important.

A PRISTINE DAY FOR FLYING

> And summer camp made it so that the aircrew and all the trades lived together for two weeks and got to know one another a little bit better. And of course, because we were going to a different place, we had to sort of live together a little bit more because you only came out normally one night a week and every other weekend and you really didn't get to know the individuals that much. At summer camp you got to know them, and fun and games, and that was a real morale builder. You were also able to do much more intensive training if you were inexperienced on type, because when you go from one aircraft type to another we may get somebody who just came out of the Regular Force into the air Reserve, but if he had

been flying maybe jets and now he's going to fly the Expeditor or the Otter or the Dakota, it's altogether different; and in order to make a person proficient and really good, the more number of hours he can get in a shorter period of time, the quicker and more thoroughly he'll learn. And the reason you can't count on a person coming out every evening and weekend is because most people have family and they deserve something from their mothers and dads who are in the Reserve. As a matter of fact, I always felt that the biggest sacrifices were made by the spouses of the people who were in the Reserves.[55]

AN OTTER EQUIPPED WITH PONTOONS, PLOUGHS THROUGH CHOPPY WATER

1964

The first quarter of 1964 brought with it significant reductions in Canada's Auxiliary air force with the disbandment of five of the eleven squadrons effective 1 April. Fortunately, 402 was not among them, although two affiliated formations – 3052 Technical Training Unit and 4003 Medical Unit – were terminated on 22 February. The "City of Winnipeg" appellation was also dropped from the title on

CONSPICUOUSLY MARKED 402 EXPEDITOR

22 February. This happened to all "City of..." squadrons, probably as a lead-in to unification. Although the "City of Winnipeg" title no longer officially applied, 402 has continued to use it as an identifier and a celebration of the unique, enduring bond between city and Squadron.

Over the next two years, the Squadron itself was as busy as ever carrying out a variety of services and duties: summer camps at Gimli and Penhold, northern operations, flying doctor services, weekly routine flights, mercy missions, and air searches for both downed civilian and military aircraft. In 1964, the Squadron received two additional Otters, which greatly increased its search and rescue capabilities. On 13 May, a pair of them flew almost seven hours in search of a pilot of a T-33 who had bailed out 17 miles northwest of Gimli. A civilian aircraft eventually found the wreckage of the T-Bird, and the pilot's body was located by the Otters and a 111 KU Albatross. During the first half of 1964, the Squadron flew a total of eighty-four routine flights.

The number of passenger miles for the period was 97,009 – 22.2% of the total flown by Auxiliary squadrons.

From 25 June to 5 July 1964, personnel of 402 took part in the annual 17 Wing summer camp at Gimli. The eight Squadron aircraft and a fifth Otter assigned to Station Armstrong were maintained in serviceable condition throughout the nine-day period. A total of 235 hours 15 minutes was flown, of which 101 hours and 20 minutes were on long-range flying exercises to maintain pilot, navigator and crewman proficiency. The longest trip was to Fredericton, New Brunswick to observe the work being done by 401 Squadron in support of the Army at Camp Gagetown and to obtain information and knowledge of joint services co-operation. The remaining hours were classified as follows: operations, including routine flights and a medical mission to stations Armstrong and Pagwa; special exercises; and other flights. The "special exercises" included a routine radiac exercise in which two Expeditors took part; "Operation Double Exposure," set up to survey and photograph minor landing fields in southern Manitoba for location, serviceability, obstructions, magnetic headings, available communications, ownership, etc. that could be used for emergency purposes; and a series of airlift, aerial field inspection, communications, and air experience flights, all carrying Army personnel in support of their annual militia training camp.

402 SQUADRON PARADING WITH WEAPONS AND LED INTO 10 HANGAR BY W/C TOM PATTERSON

OTTER TECHNICIANS CLASS GRADUATION PHOTO

As at previous summer camps, much emphasis was placed on the training of groundcrew, notably in the mechanical and clerical trades. The practical aspect was emphasized, and all Auxiliary airmen received forty-eight hours of training during the summer camp. Auxiliary personnel carried out all first-line servicing and rectified approximately 40% of the unserviceabilities. The photo recon and Army co-ops were also supported by Auxiliary groundcrew. Also on the agenda were familiarization and continuation training of aircrew and groundcrew of airborne radiac equipment, and groundcrew in the handling of ground-to-air telecommunication and radiac transmitters. This simulated training gave aircrew personnel experience in radiation monitoring.

Mercy flights were also logged by the Winnipeg Bears in 1964: in September an Otter carried out an emergency evacuation of infant twins born prematurely to the wife of an airman at Station Gypsumville. In October, an Expeditor delivered a shipment of blood to Dauphin for a boy with haemophilia. Also in October, Expeditors and Otters of 418 Squadron (Aux) Edmonton were flown to Winnipeg as an exercise in emergency deployment and as a test of the servicing facilities of 402 Squadron. On the afternoon of the 17th, 402 and 418 squadrons supplied an Expeditor and an Otter to take part in the annual Auxiliary flying competition. At the close of the competition, W/C D.R. Scott, CO of 402, presented the Auxiliary trophy to the Commanding Officer of 418 Squadron.

1965

Anno domini 1965 was a memorable one for Canada because it was in that year that the country adopted its new flag. The eleven-point, stylized maple leaf that graced the centre field of the flag was adopted for the roundels applied to military aircraft, and the new flag replaced the Canadian Ensign that for several years had appeared on their tail surfaces. By early April, these changes in livery were being applied to 402's aircraft. For the Squadron in particular, excellent northern experience was being gained by Otter air and groundcrews. Airlift flights of Army Signals Corps inspection personnel to Nuclear Defence Fallout Recording Sites commenced in late January and continued on into the third week of February. Locations visited included Island Lake, Moose Lake, Dauphin River, Brochet, Gods Lake, Gods River Narrows, Norway House and Lynn Lake. With so many missions now taking place in the North, the need for specialized winter training was answered with the inauguration of "Operation Snowshoe": between 8 and 14 February, 402 and 17 Wing, equipped with two Otters, participated in a special operation at Snowshoe Lake forty miles northeast of Lac du Bonnet. The objective was to give pilots and groundcrew experience in operating the Otters from frozen lakes, and maintaining them in cold weather. Equipment was flown to the site in advance of the exercise, and two para-rescue airmen from 111 KU also arrived early to assist in preparations. The Otter, as expected, proved itself to be trustworthy and reliable; no serious mechanical problems were encountered following nights during which the aircraft were exposed to -40°C temperatures. The crews gained valuable experience (and exercise) installing wing covers and operating catalytic heaters.[56]

AN LAC GETS SOME BASIC ELECTRICAL POINTERS FROM THE "FLIGHT"

AN OTTER ON SKIS

By the third week in March, Squadron flying activity was running at a high level due to revised allowable rates. Most aircrews had completed quarterly requirements (30 hours in Air Transport Command) and the keeners were doing additional long-range trips and completing night-flying requirements as well. With winter's end in sight, the Otters would soon be shedding their skis and consequently fewer "airports" would be available for those who enjoyed the freedom and challenge of operating in the outlying areas.[57] Meanwhile, the Otter pilots were busy compiling weight and balance statistics for the ever-increasing number of cargo flights being assigned to the Squadron. Many tons of supplies were being delivered to Portage, Rivers and Gimli, and future plans called for the broadening of the mandate to include the Pinetree Line stations at Beausejour, Sioux Lookout and Gypsumville.[58]

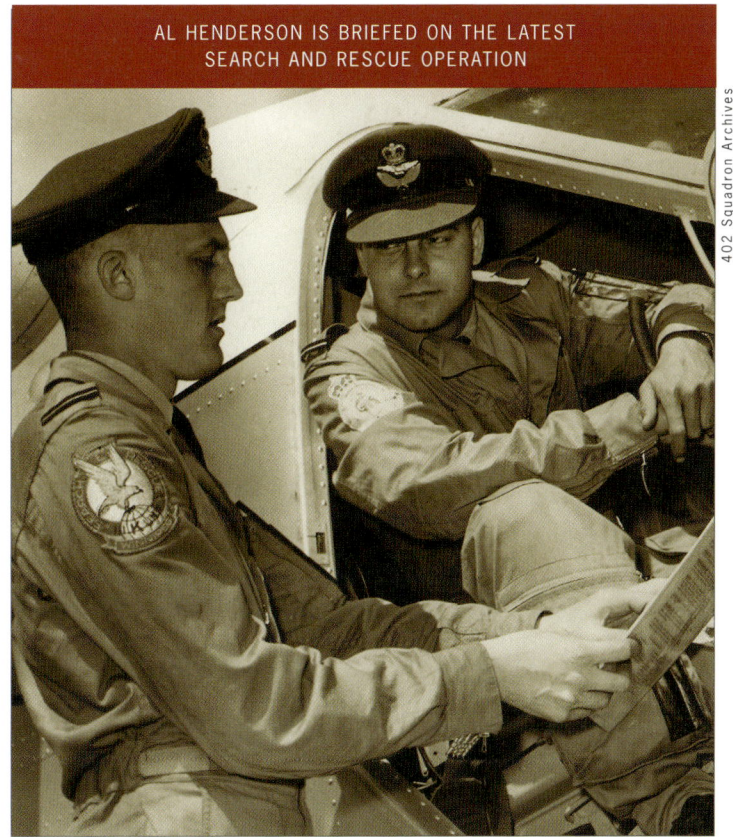

AL HENDERSON IS BRIEFED ON THE LATEST SEARCH AND RESCUE OPERATION

A successful rescue was accomplished on 22 May when three Otters went searching for a civilian Stinson that had gone missing in northern Ontario. It was found overturned in a lake 17 miles from its destination with all three survivors safely on shore. A 111 KU Albatross subsequently rescued the three. 57 Auxiliary and 31 Regular Force personnel were front and centre at the 1965 summer camp held at RCAF Station Penhold, where three Otters and as many Expeditors were based for the two-week period. A fourth Otter was deployed to Camp Shilo in the first week to transport Army personnel, drop supplies, and carry out reconnaissance and photographic work. All told, Squadron aircraft flew 349 hours and 45 minutes, including routine flights, in the three-week period, and the Shilo-based Otter carried 152 Army personnel. In one exercise, it successfully dropped a forward medical aid station, including tents and surgical supplies and instruments from an altitude of 250 feet. One new feature of the 1965 camp was a ten-day course in basic survival given at Ram River, ninety miles west of Penhold, by an instructor from the Edmonton-based Survival Training School. Closer to home, a Squadron Otter was stationed at Flin Flon to provide transport and to evacuate Militia personnel engaged in a bridge-building exercise at Snow Lake.

AN ENGINE START ON A 402 EXPEDITOR

On 10 October, the public got a chance to check out the 402 Otters at a civilian air show at Carman; after forming part of the static display, the aircraft demonstrated a short-field take-off and a low-speed flypast. W/C Scott was appointed Chief Administration Officer (CAdO) at 17 Wing and was succeeded as CO by W/C J.A. Brown, CD: the official change-of-command took place on the 21st of the month at a Wing parade. Meanwhile, a Squadron Otter was withdrawn for United Nations Observer Force duty in Pakistan with the Canadian personnel assigned to the United Nations India-Pakistan Observer Mission (UNIPOM). Two other Otters took part in an exercise at Beausejour in which six crewmen, who had previously attended lectures arranged by Command Headquarters, became the Squadron's first Auxiliary personnel to qualify in supply dropping. In December, the Winnipeg Bears inaugurated "Operation Santa Claus" to bring Christmas cheer to isolated northern points. Eaton's department store supplied the presents that year, and 402 delivered the gifts to thirty-seven children at Jackhead Harbour on Lake Winnipeg. The success of "Op Santa" ensured its continuance for many years.

By the time 1965 had drawn to a close, 402 had logged 166,862 passenger miles and had carried 19,215 ton-miles of cargo. These impressive figures in fact represented 23.6% and 25.9% of the respective totals for all six of the Auxiliary squadrons. In addition, the Squadron flew over ninety-six search and rescue hours.

1966

402 Squadron was proving to be something of a godsend to many remote northern communities. An example was played out in January 1966, when toys and school supplies donated by the Red River Chapter of the Imperial Order of the Daughters of the Empire were airlifted to Warren Landing School. The overnight "Operation Snowshoe" that had been developed the year previous utilized three Otters in January to provide eleven pilots and fifteen ground servicing personnel with training in winter flying, servicing aircraft

SGT BOB JAMES (LEFT) CROSSES THE TARMAC AT WINNIPEG IN THE EARLY 1960s. 30 YEARS LATER, HE BECAME THE SQUADRON'S HONOURARY COLONEL

in extreme weather, and basic northern survival. Local personnel and a representative of 111 KU, who demonstrated survival techniques, also provided instruction. The above-mentioned aircraft that had been dispatched to Pakistan was replaced, and from 5 to 7 February, 402 participated along with Toronto, Montreal and Edmonton Auxiliary squadrons in "Operation Totem II". The session included parachute and free-drop training, and competitions between the squadrons. The month of February brought with it the inauguration of the "Operation Airlift" series of operations exercises in support of the Army. Conducted throughout the course of 1966, these flights involved carrying Army personnel from a number of units principally to Camp Shilo but to other sites within and outside the province, to conduct their range exercises.

Also on the Army's behalf was the Squadron's involvement in "Operation Sneak Peek" in May, again at Shilo, when six transport movements, seven supply flights and drops, and seven visual and photo reconnaissance sorties were flown. During these Army co-ops, Otters and/or Expeditors, as appropriate, were deployed. Five Otters were laid on in early May for training and check paradrops at Beausejour, where five crewmen were checked as qualified Otter paradrop dispatchers. However before all that happened, the Winnipeg Bears were called upon in the second week of April to assist in dealing with the rising waters of the Red River. This time, local press and National Film Board parties were airlifted on a flood survey flight by an Otter, and later on eight officers and thirty-six other ranks assisted provincial authorities in flood control duties. By January 1966, changes in Canada's military establishment had seen 402's home base renamed "CFB Winnipeg". Nor was the year without its usual involvement in search and rescue activities. "SAR Green" was a successful operation wherein an Otter was not only involved in the finding of a northern crash site, but in transporting an RCMP constable and a coroner from Norway House to the scene.

EXPEDITORS LINED UP OUTSIDE 10 HANGAR WINNIPEG
MGen E.I. Patrick

It was all good news when the Air Transport Command Aircrew Standards Unit made its annual visit to the Squadron in late June. No major operational deficiencies were noted, and 402 was assessed as maintaining a good standard of operating proficiency. Then it was off to CFB Toronto for summer camp, in which thirty-six Auxiliary and twenty-one Regular Force personnel took part. Two Expeditors and three Otters were based at Toronto for training, transport, category checks and support of Army exercises. Two Expeditors remained behind in Winnipeg to cover RF 4402[59] and the remaining Otter was deployed to Camp Shilo for transport and aerial recce. Twenty-one personnel and 1230 pounds of cargo were carried by the Toronto-based aircraft, and night supply drops to Army units in the field were carried out for the first time. The Expeditors and Otters flew a total of 240.55 hours, with much of the above flying having been in support of the Army exercise "Tempered and Polished Bayonets". Meanwhile, the time logged by the Otter deployed at Shilo included air supply drops to troops in the field and aerial photo and visual reconnaissance, as well as transport and familiarization flights.

The final quarter of 1966 found five aircrew and eleven ground personnel back at CFB Toronto, this time for a paradrop competition. The Winnipeg Bears, using Toronto Auxiliary squadron aircraft, were awarded second place in the contest. Another landmark date was 1 November, when 402 bade farewell to the venerable Expeditor. The aircraft were now to be ferried to the Air Force's Saskatoon storage facility by unit pilots, as and when time allowed. The last one was finally dispatched on the 16th of the month. The Squadron was now equipped with five Otters in wheel-ski configuration. In the meantime, the Otters flew three airlift sorties to MacDonald, Manitoba where members of the 3rd Regiment, Royal Canadian Horse Artillery commenced their escape and evasion exercises. This year, "Operation Santa Claus" visited Island Lake, Manitoba with gifts donated by the Great West Life Assurance Company.

1967

The year 1967 brought with it the anticipated share of air searches and training courses that were by now standard fare for 402. Summer camp was held at RCAF Station Penhold where duties included Air Cadet familiarization and transport. The summer months of July and August also saw the Squadron transporting Army cadets from Fort William and Atikokan, Ontario to Clear Lake on Riding Mountain for their summer camp. Certainly the most engaging experience during the year would be the Canada Centennial celebrations, which would be, by definition, unique in the country's history. Participation in public events, a form of activity in which the Squadron had a long tradition, was an obvious way of taking part. During the first weekend of July, fly-ins were carried out in Ontario and Saskatchewan: W/C Patterson piloted an Otter to Dryden where the local flying club had scheduled a ground and air display of aircraft, and F/L Demare demonstrated the capabilities of his Otter with short take-offs and landings at the Moosomin Flying Club. Not to be outdone was the acclaimed 402 City of Winnipeg Squadron Pipe and Drum Band: the Canadian Forces in Manitoba had their own Centennial Caravan, comprising open-air displays and a pair of large vans that showcased the three Services. Under the control of Training Command HQ, Winnipeg, the caravan visited 21 communities in the province between 22 June and 22 July, and the band played at seven locations.

1968

On 1 February 1968, the Royal Canadian Air Force was abolished as a legal entity, and 402 was integrated into the Canadian Armed Forces. Thus closed a memorable chapter in the Squadron's history. A chapter that had seen 402 go through numerous significant changes, including seven new designations, six aircraft types, nine commanding officers and five higher formation moves. Indeed, "Adaptable" might well have been chosen as the Squadron's motto, for as forceful as the winds of change raged during the '50s and '60s; they never blew harder than during the birth of the Canadian Armed Forces. Even as the passing of the beloved RCAF was being mourned, the "green" air force began to emerge, in many ways, vastly different than its predecessor. As always, the Winnipeg Bears rose to the challenge.

NOTES

1   Hitchins, F.H., "No. 402 (Fighter) Squadron". *The Roundel* 2(5): 31-35, 1950. p. 34.

2   The (FB) designation stands for "Fighter Bomber".

3   Organization Order 727, Formation of 9402 RCAF Detachment, Winnipeg, Man., 22 May 1946. p. 1.

4   Greenhous, B. and H. Halliday, *Canada's Air Forces 1914-1999*. Art Global, Montreal, 1999. p. 122; Bristol Aerospace Ltd, *50 Years of Technology, Vol. One: The MacDonald Era, The First Quarter Century.* Bristol Aerospace Ltd. Winnipeg, 1980.

5   W/C M. Reid RCAF (Ret'd), personal communication, 2005.

6   "Plan E", Post War Plan for the Royal Canadian Air Force, 1948-49. Appendix A, Sheet 1 and 5. Quoted in Joost, M., "The RCAF Auxiliary and the Air Defence of North America, 1948 to 1960". Canada in NORAD: 7th Annual Air Force Historical Proceedings, Peterson AFB, Colorado Springs, June 4 to 9, 2001. p. 86.

7   "Pukka": Hindi word for good, proper, reliable; "gen": information– thus, "reliable information". Langeste, Tom. *Words on the Wing*, p. 223, .

8   The records are inconsistent as to the number of Vampires TOS by the Squadron. The Daily Diary for 19 April 1948 notes that the first Vampire arrived this date. Then, "F/L Dempster of Aux Support Unit and two pilots from Montreal Aux Sqn arrived Friday 25 June 48 with three Vampire Jet aircraft. This completes the Sqn complement of four jet aircraft." However, the log for 24 July 1948 reads in part: "S/L Hudson arrived in the late afternoon with the Sqn's fifth Vampire." Thus, *five* a/c are accounted for in this sequence of statements. The August 1948 Historical Record makes reference to five Vampires having City of Winnipeg crests painted on their noses.

9   Searchlight co-ops were practice sessions carried out to the mutual benefit of both Army and Air Force participants. The soldiers attempted to find the aircraft and hold it with their searchlights, while the aircraft tried to avoid being caught and held in the searchlight beams.

10   G/C D. Gray RCAF (Ret'd), 402 Interview Project, Summer 1987.

11   Report of the Department of National Defence for the Fiscal Year Ending March 31, 1950. p. 61.

12   Report of the Department of National Defence for the Fiscal Year Ending March 31, 1950. p. 64.

13   "Cab rank support" involved the deployment of fighter-bombers in a holding pattern that enabled them to be called down in sequence via radio to provide immediate firepower support to ground troops.

14   Vaughan, A.P., *418 City of Edmonton Squadron History*. The Hangar Bookshelf, 1984. pp. 74-75.

15 McNorgan, P. D. and R.W. Patrick, *402 City of Winnipeg Squadron History 70th Anniversary Edition*. Winnipeg, 2002. p. 28.

16 The sobriquet "Bugsmasher" refers to the large windscreen on the Expeditor. Frequently during the summer months, upon landing, they were covered wuth the remains of "bugs".

17 Morriss, W., "'Bullet-Ridden' Newsmen Undaunted, They Want Their Coffee Anyway". *Winnipeg Free Press*, 7 October 1949. p. 11.

18 Young, D., "'Enemy' Down In Flames: RCAF Plays Battle Games". *Winnipeg Free Press*, 11 July 1950. p. 8.

19 Vaughan, p. 79.

20 Interview with W/C Thomas Patterson RCAF (Ret'd) by Norman Malayney.

21 Davis, L., *Air War Over Korea: A Pictorial Record*. Squadron/Signal Publications. Carrollton, Texas, 1982.

22 This designation remained in effect until 1 September 1951.

23 Interview with W/C Thomas Patterson RCAF (Ret'd) by Norman Malayney.

24 Quoted in Hatch, F.J., "Salute to the Auxiliary". *The Roundel*, 1964, Vol. 16, No. 2. p. 12.

25 Martin, P. with J. Griffin, *Royal Canadian Air Force Aircraft Finish and Markings 1947-1968*. Privately published, 2003.

26 Quoted in Hatch, p. 12.

27 402 Sqn - an interview with W/C Thomas Patterson, RCAF (Ret.), Interview by Norman Malayney.

28 Queale, L.W., "Exercise Nugget". *Voxair*, Winnipeg, 22 August 1952. p. 9.

29 F/S Maurice Munn RCAF (Ret'd), 402 Interview Project, Summer 1987.

30 G/C Neil Scott RCAF (Ret'd), 402 Interview Project, Summer 1987.

31 Interview with W/C Thomas Patterson RCAF (Ret'd) by Norman Malayney.

32 A "pipeline" was a ground-controlled intercept (GCI)/ground-controlled approach (GCA) landing procedure, in which a radar station handed over control of an aircraft directly to the GCA controller at the destination airport.

33 The "402 Pilots' Diary – Summer Camp 1953" reads in part: "Fri 10 Jul … In afternoon 3 Mustangs were ferried to Bagotville to take part in Exercise Tailwind. Pilots were returned by Beechcraft Friday night", implying that 402 aircraft, but not personnel, were involved in the operation.

34 Joost, p. 92.

35 Written communication to R. Lanoway, 2004..

36 Written communication to R. Lanoway, 2004.

37 Baglow, *Canucks Unlimited: Royal Canadian Air Force CF-100 Squadrons and Aircraft, 1952-1963*. Canuck Publications. Ottawa, 1985. p. 29.

38 Interview with W/C Thomas Patterson RCAF (Ret'd) by Norman Malayney.

39 McNorgan and Patrick, p. 33.

40 Joost, p. 92.

41 Pickett, J., *Into the Sausage Machine: The History of 22 Wing*. North Bay, 1994. pp. 67, 75, 77.

42 J.M. Reid, W/C, CO 402 (F) Sqn (Aux), Narrative Report, Historical Record, September 1954.

43 Kostenuk, S. and J. Griffin, "RCAF Squadron Histories and Aircraft 1924-1968." *National Museums of Canada, National Museum of Man, Canadian War Museum Historical Publication* 4, Ottawa, 1977. p. 147.

44 Interview with W/C Thomas Patterson RCAF (Ret'd) with Norman Malayney.

45 They were still there in 1961 awaiting disposal by Crown Assets

46 Fletcher, D. and D. MacPhail, *Harvard! The North American Trainers in Canada*. DCF Flying Books. San Josef, B.C., 1990. pp. 173-202.

47 Quoted in Hatch, p. 13.

48 Historical record, 402 Sqn (Aux) Winnipeg, Manitoba, Period from 1 Dec 1959 to 31 May 1960, p. 4.

49 Col. Ernest Harris CF (Ret'd), 402 Interview Project, Summer, 1987.

50 Col. Ernest Harris CF (Ret'd), 402 Interview Project, Summer, 1978.

51 Milberry, L., *Sixty Years: The RCAF and CF Air Command 1924-1984*. CANAV Books, Toronto, 1984. p. 341.

52 Nicks, D., J. Bradley and C. Charland, *A History of the Air Defence of Canada.* The NBC Group, Ottawa, 1997. p. 121.

53 Capt. Al Henderson CF (Ret'd), 402 Interview Project, Summer, 1987.

54 "radiac" is an acronym for "radioactivity detection, indication and computation" and refers to activity and equipment directed toward detecting and measuring nuclear radiation intensity.

55 Col. Ernest Harris CF (Ret'd), 402 Interview Project, Summer, 1987.

56 Gittel, A., "402 Sqdn. (Aux)". *Voxair*, March 19, 1965. p. 6.

57 Gittel, A., "402 Sqdn (Aux)". *Voxair*, March 19, 1965. p. 6.

58 Gittel, A., "402 Squadron News". *Voxair*, April 2, 1965. p. 7.

59 RF 4402 referred to regularly-scheduled flights assigned to transport squadrons. Auxiliary squadrons were often feeders for military service flights operated by the Regular Force. The latter conducted the long-range flights, while the Auxiliary (and some Reg Force squadrons) would do the short hauls (M. Joost, personal communication, 2005).

Flight Sergeant Minto

402 Squadron's mascot is a grizzly bear named Minto, who became a premier attraction at the Winnipeg Zoo during the 1950s and '60s.

Born in 1950, the bear was named after the town of Minto in the Yukon, close to where he was born. As a young cub, he hung around a construction company work site, before finally being caged and flown south to the Calgary Zoological Society in the "Stampede City". After a short stay in Calgary, he was given to the Winnipeg Zoo, arriving by train on 31 January 1951.

Minto was given the rank of Sergeant, and received his appointment as official mascot at a luncheon in Winnipeg. His face became the model for the unofficial squadron badge, a replica of which the Commanding Officer, W/C W.B. Breckon, DFC, presented the city. The badge eventually found its way onto the bars of Minto's cage.

Sgt Minto was married with children. His wife, Maude, delivered their first-born cub on 8 February 1954. The baby had the distinction of being the first grizzly bear born in captivity in Canada.

Although Minto became a FSgt in 1958, he never made it onto the official 402 Squadron badge. The College of Arms rejected 402's proposal drawing of a grizzly bear head, saying it bore too much resemblance to that of a wolf. At

FSGT MINTO GREETING VISITORS AT THE WINNIPEG ZOO

this time, the Squadron decided to adopt the current "totem" bear as an alternative. The earlier 402 Mustang scheme featured Minto's profile within a cannon ball, as standard "nose art" on most of the aircraft. The "totem" bear first appeared on some of 402's Expeditors in the mid to late 1950s.

FSgt Minto soldiered on into the '60s as the Squadron mascot, before eventually being sent to the Chicago zoo. In fact all the bears were farmed out as the Assiniboine Park Zoo constructed new enclosures for the bruins and had no place to house them in the interim.

Minto frequently appeared on the cover of the 402 newspaper, ironically titled Totem Talk. The grizzly bear seemed to have fallen out of fashion until the 1990s, when "Grizzly" replaced "Gonzo" as the Dash-8 call sign. Minto's image now lives on, adorning most of the Squadron's room titles in 16 Hangar, and as proposed tail art for the "blue" scheme CT-142s.

Ranks

Equivalent ranks: RCAF to CF

Officer Ranks

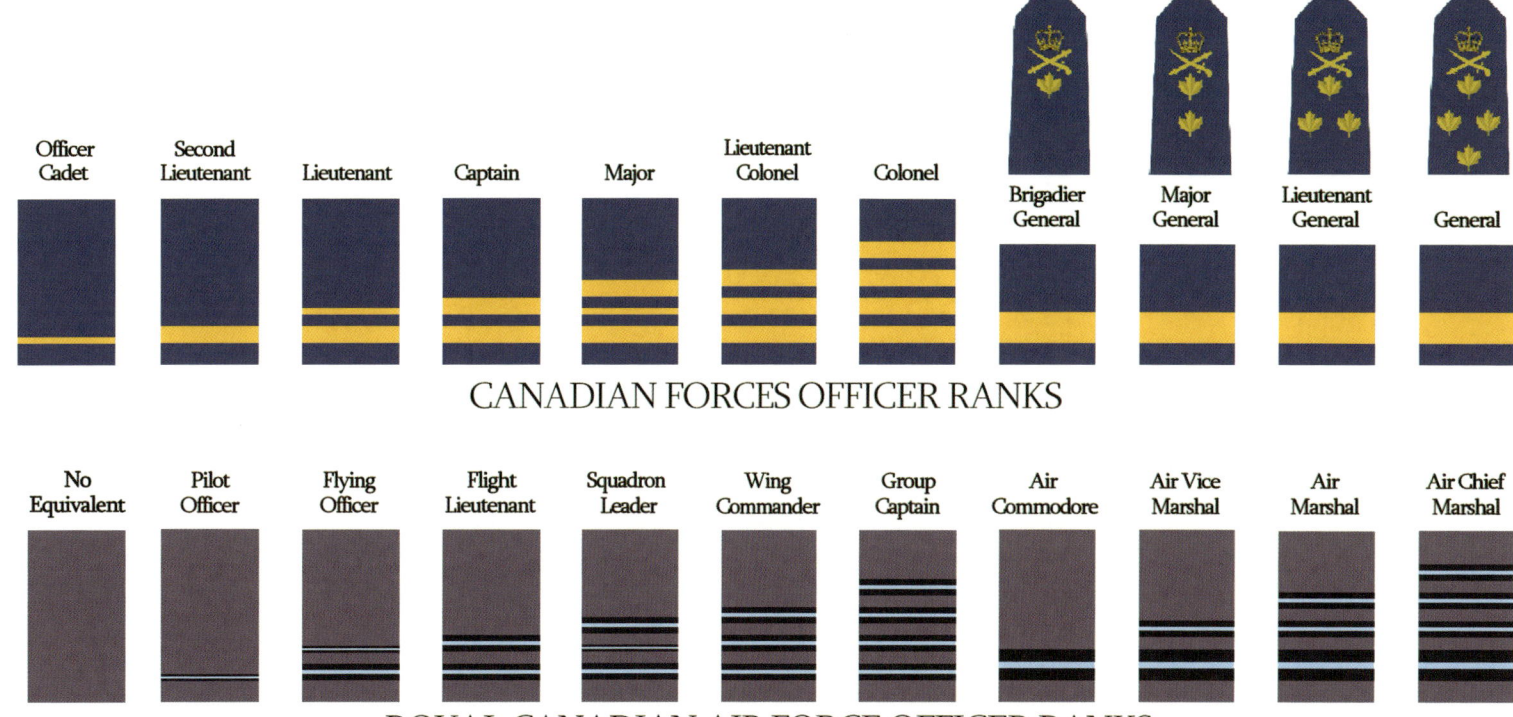

CANADIAN FORCES OFFICER RANKS

ROYAL CANADIAN AIR FORCE OFFICER RANKS

RANKS

Equivalent ranks: RCAF to CF

Other Ranks

CANADIAN FORCES NCM RANKS

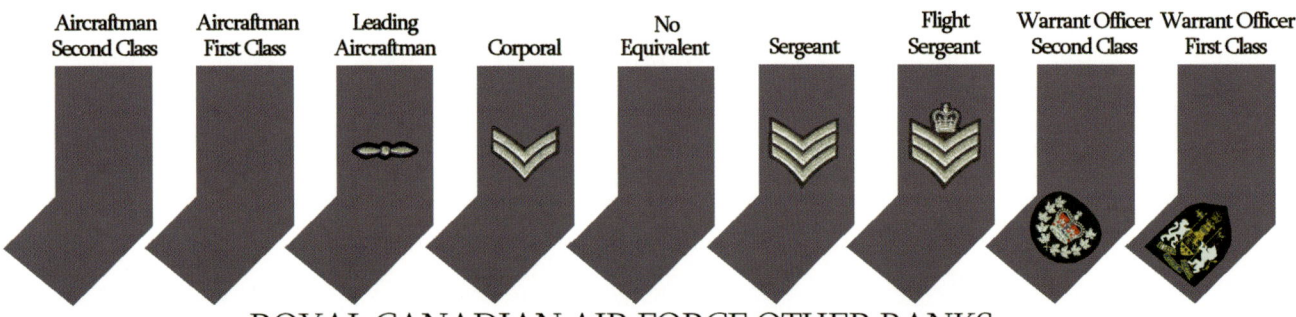

ROYAL CANADIAN AIR FORCE OTHER RANKS

CT-142 803 PROFILE WITH 75TH ANNIVERSARY MARKING STRIPES ON RUDDER, SPINNER AND "CITY OF WINNIPEG SQDN" IN SCRIPT ON OUTBOARD COWLINGS Art by Pat McNorgan

Post-Unification

Lieutenant-Colonel Dean C. Black

In the post-unification and integration period of the late 1960s and early 1970s it is relatively accurate to describe the role of 402 Squadron, as a Reserve Force member of Canada's air force responsible for medium transport, VIP executive transport, and search-and-rescue. For much of the closing decades of the Cold War this remained so. Toward the end of the Cold War however, a change in the role was realised that reflected in part the increasing technical sophistication of air operations, specifically air navigation training and a capability-based preference for greater effectiveness in flight operations. In September 1958, USAF historians were well informed with respect to the marvels of enhanced effectiveness in air force operations. "Only one thing more fantastic than modern air power," they said. "That is…tomorrow's airpower." The story of 402 Squadron therefore is typical of successful air force units in the post-Second World War period. Success, it is said, follows from experience and experience, from good judgement. 402 Squadron cut its teeth in the crucible of fighter pilot combat, but enjoyed longevity in post-Second World War air operations, owing to the sound judgement of air force decision-making that respected both the importance of air force support missions, and the role of Reserve personnel in modern-day military forces. The chronological representation of 402 Squadron's postwar activities that follows, therefore, is based on these themes reflecting two fundamental elements of air power: how essential to a nation's airpower is support and airlift; and how critical to postwar air forces have the Reserves been? The former is considered one of four primary elements of air power that directly associated with the nation's air force. The latter is considered to be an important aspect of one of five secondary elements of air power; namely, the nation's people.

A WINTER SCENE FROM 1974 WITH AN OTTER TAXIING AWAY FROM 16 HANGAR. THE AIRCRAFT CLOSEST TO THE CAMERA CARRIES THE WINNIPEG CENTENNIAL CREST ON THE PILOT'S DOOR

In some respects, the historical information pertaining to 402 Squadron presented below may not seem much different from that of any other air force squadron, but this is precisely the point. While articulating his vision for America's 1960s/70s space program, President John Fitzgerald Kennedy once said "We do these things not because they are easy, but because they are hard…"[1] Like many before him and many that have followed, Kennedy acknowledged the dangers associated with flying operations of the time. It is as if he knew injury and death would be companions to those who risk their lives establishing the nation's air power. The history of 402 Squadron is thus not very different at all from other squadrons that suffered the loss of remarkable individuals who perished in accidents from time to time.

F/O Vernon Bastable was one such officer lost to 402 Squadron on 27 March 1949, and of whom Squadron personnel continue to be reminded. Bastable was on a training flight when his Vampire 17032 crashed in the Winnipeg suburb of Charleswood.

Bastable had been awarded the Military Cross in 1948 for his past heroic work during the war with the Czech resistance. After escaping from a prisoner of war camp with a Czechoslovakian friend, he opted to stay and fight with the resistance. Responsible for small arms training, he led many raids against enemy facilities and supply lines. Posthumously promoted to the rank of Flight Lieutenant, he continues to be remembered via the Vernon Bastable Memorial Award presented to 402 Squadron's Regular Force Airman/Airwoman of the year. This award recognizes a Regular Force squadron junior Non-commissioned member who displays outstanding qualities in trade skills, leadership, deportment and dress.

Why reiterate Bastable's exploits, and his relevance to the modern-day 402 Squadron? Vernon Bastable is an icon today because his loss was a significant one. Events such as this tragedy tend to have a lasting effect, shaping the history that follows. At about the time Bastable was waging war in Czechoslovakia against the Nazi regime, a French army Captain and soon-to-be one of the world's foremost historians named Fernand Braudel was under lock and key, he too having been arrested by the enemy. Today, Braudel is remembered for his publication *On History*, and more specifically, his method for articulating history. Braudel proposed the *longue durée* as a complement to the other type of historical method; namely, *l'histoire événementielle*. To Braudel it was important to provide the somewhat stable, longer duration and overarching context to the turbulent and visible and more frequently changing events of the day. It is to this method that the following narrative ascribes. The longue durée in this history of 402 Squadron addresses the constant themes of training, preparation and support, while *l'histoire événementielle* must include events such as the passing of Vernon Bastable, for they serve as a lens through which one can measure the Squadron activities and challenges that followed.

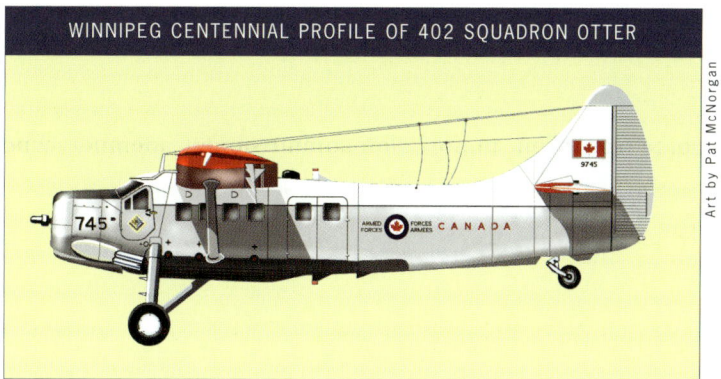

One of the most notable everyday changes that integration of the three services brought to the air element involved the loss of RCAF rank names. Thus an air marshal became a lieutenant-general, a squadron leader a major, a flight sergeant a warrant officer, and so on.[2]

In October 1971, members of 402 Air Reserve Squadron undertook to improve their map reading, direction finding and general command and control operations training. The Squadron had recently bid farewell to their outgoing Commanding Officer Lieutenant-Colonel (LCol) E.J. Harris, and welcomed the incoming LCol R.D. Wilson.

In so far as training is concerned however, under the auspices of "Exercise Windmill I and II," Captains (Capts) J. Reeve, B. Weber, M. Massier, R. Salome, R. Scott and Lieutenant (Lt) D. Bailey brought Squadron personnel through a number of training activities as part of an annual refresher in basic field skills and preparations. Exercise Yellow Quill followed in 1972, and for all intents and purposes provided the Squadron with a venue to conduct training very much similar to that provided during the "Windmill" series of exercises.

FORMER 402 CO COL ERNIE HARRIS, MIDDLE, HAS COL MOE GATES ALMOST DOUBLED OVER IN LAUGHTER WHILE BGEN THORNYCROFT, LEFT, THINKS IT OVER

On 15 July 1972, both 402 and 418 Squadrons wrapped up their "Summer Concentrations." Activities involved a number of flights throughout the Yukon and Northwest Territories. These exercises served a number of purposes, but their main focus was to expose aircrew to flight operations in the region, and to respond to the needs of Northern Region Headquarters (NRHQ). While the base of operations was Yellowknife, aircraft and crews operated from Whitehorse and Inuvik as well.

A LINE OF 402 DAKOTAS AT CFB SUMMERSIDE, 1979

The post-operation report was clear on the benefits of these exercises, as expressed by Commanding Officer LCol E.T. Wagner: "Squadron personnel obtained invaluable experience in operating in Northern Canada, both at and away from a fixed base, and useful information and photographs were provided to NRHQ" (Northern Region Headquarters). The 1972 Combined Concentration operated successfully under the wing concept, but it was the Commanding Officer's opinion that similar results might have been achievable using separate concentrations "since no [single] operation was beyond the capability of any [single] squadron." LCol Wagner concluded "Under this system, the presence of the military in the North would be extended over a longer period." The question of Canadian sovereignty (read military presence) in the North is not a new one, and some far-sighted individuals were voicing their concerns. Colin S. Gray, Executive Secretary Strategic and International Studies Commission, Canadian Institute of International Affairs, Toronto explained, "…Canada's 'sovereignty' might need military protection in the years ahead. Canadian territory, the territorial sea, and the seabed and finally, Canadian airspace are all reviewed as areas of actual and potential military endeavour."[3] From this, it should be clear why almost without exception every Canadian Forces unit had a vested interest in maintaining proficiency at operating in Canada's northern regions. Success of a single flight or trek across the country's vast largely uninhabited regions to the north depended on the collective efforts of many talented and disciplined individuals. The 402 Squadron record for successful operations in the north is both diverse and courageous.

The following year, an "Air Reserve Summer Concentration" operated out of Canadian Forces Base (CFB) Edmonton from 30 June to 8 July 1973. Personnel from 3 Air Reserve Region Headquarters, 402 Squadron and 3 RSU participated. Capt R.A. Scott was appointed coordinator. According to Capt Scott's draft orders, Colonel (Col) E.J. Harris was to be joined by 20 other officers and 17 non-commissioned officers. Their mission was to "train in the operation of a field deployment camp."

In 1974, the summer concentration was similar to the 1972 excursion. Operating out of Yellowknife from 29 June to 7 July, the Squadron enjoyed support from 440 Squadron and its facilities. A total of 56 personnel from 402 Air Reserve Squadron, 3 ARWHQ, 3 RSU and CFB Winnipeg deployed to the Yellowknife airport and operated for the most part out of Akaitcho Hall. The detachment's supply section, meanwhile, operated out of the RCMP hangar facility.

The Squadron's important work in the north seemed to go on despite the tragic loss of Capt Jack Reeve earlier that year. On 19 May 1974 Capt Reeve was killed when an Otter aircraft in which he was flying, crashed during an exercise at Swallow Lake Airstrip in Ontario. The *Globe and Mail* reported the next day that "… [t]he crash of a Canadian Forces single-engined Otter aircraft near Swallow Lake, about 40 miles west of Thunder Bay [took] the life of Captain John Ernest Reeve of 402 Air Reserve Squadron Winnipeg." In response to the devastating loss of the 41-year-old airman, the "City of Winnipeg Reserve Air Force Association" instituted an award late in 1974 to honour Reeve's memory. The Jack Reeve Memorial Award recognizes a Reserve Squadron member who displays outstanding qualities in trade skills, leadership, deportment and dress. Captain Reeve's widow, Jackie Hall presents the award annually at a ceremonial dinner.

LONG-TIME 402 SQUADRON AND ASSOCIATION MEMBER, CPL TULSE DAS TAKES RECEIPT OF THE 1980 JACK REEVE MEMORIAL AWARD

THE FUNERAL OF CAPT JACK REEVE

CHAPTER 4

Sometime after the 1974 summer concentration concluded, 402 Air Reserve Squadron, or at least a select few members of the Squadron, embarked on a special training session identified as "Exercise Panti-Hose". A post-exercise report written sometime after February 1975 explained the unusual name for the exercise: "The exercise was manned 100% by female personnel." The women were trained in basic survival techniques and actions, as well as administration relevant to setting up such a training exercise. If it were not for the enthusiasm and wholehearted co-operation provided by the two survival instructors; Lt D.C. Houston, 402 ARS and Corporal (Cpl) G. Treen, 3 RSU, the exercise would not have been the success it was.

It seems that with each passing year the summer concentrations grew, both in numbers of personnel attending and in terms of the improvements to facilities. In 1978, semi-permanent messing facilities (each for Officers and NCOs) were established with standard mess protocols and regulations applying. In 1979 however, an Air Reserve Summer Deploy-ment exercise titled "Abegweit I" took place at Summerside, Prince Edward Island, from 14-28 July. LCol R.W. Slaughter, CD, acted as Detachment Commander. Four Dakota aircraft deployed; the first as Advance Party, while the three remaining Dakotas ferried the initial part of the main body. Thirty-one personnel deployed on the four Dakota aircraft, while the balance of the participating personnel, thirty-three more, arrived in Summerside aboard a Hercules aircraft.

Perhaps the most interesting event to affect 402 Squadron in 1980 was the amalgamation of the Squadron with 3 Air

402'S LT NEIL KEATING SHOWS NAVY AND ARMY RESERVISTS THE OFFICE OF A T-33

Reserve Wing. LCol R. W. Slaughter, Commanding Officer 402 Squadron, assumed command of the larger unit on 18 November 1980 after a short trial period was deemed a success.

On Sunday, 22 November 1981, it was reported that two 402 Dakotas, commanded by LCol Malcolm Joyce and Major (Maj) Ron Clayton, augmented the efforts of two Buffalo aircraft and five helicopters from CFB Trenton's 424 Squadron and CFB Petawawa 427 Squadron, as well as four private aircraft – all in an effort to find a missing pilot and his airplane reported lost somewhere between Fort Hope and Thunder Bay. More than thirty trained volunteers from the Lakehead Search and Rescue Unit, a private citizens' organization, worked as spotters in an attempt to locate the pilot – Pooran Shivcharan. Flying a twin-engine Aztec, Shivcharan by this time had been missing for ten days.

In 1982, LCol M.S. Joyce, Majs B. Cunniff, L.E. Olson and R.B. Ainley, Capt D.W. McDonald, Sgts J.G. Dupuis, P.W. Schmitt and B.P. DesAutels, and Cpl J.N. Ferwerda all earned DC-3 Dakota 1,000 hour plaques. Cpl Johannes Ferwerda, a technical crewman who served with the Squadron during the entire 402 Dakota period, describes the award and his fondness for the "Gooney Bird"[4]:

"It (the 1,000 Hour Plaque) meant a lot to me because of the aircraft. Being from a Dutch immigrant family, and knowing what a big part the Dakota played in the liberation of the Netherlands, made it very special for me. That 1,000 Hour Plaque still hangs in my home and remains a prized possession"[5]

HONOURARY COLONEL
PEARL MCGONIGLE

The 50th Anniversary Reunion Committee was also struck with Maj Maryalyce Massier in charge. A number of VIP trips took place with Her Excellency the Honourable Pearl McGonigle, CM, OM, leading the way with ferry flights to northern Manitoba in support of her Levee tours.

On 8 May 1982, ten members of 402 and a representative of the Western Canada Aviation Museum (WACM) departed Winnipeg for Orlando, Florida. Their mission: to return to Winnipeg at the controls of a Junkers JU 52/3m. The trophy they were bringing home was actually a Casa 352; coincidentally, the aircraft had been both constructed in Spain and used in the Spanish Civil War. Upon arrival in Florida, two Confederate Air Force colonels, Dick and Dale Markgraf – a father and son team, arrived from Colorado to check the Canadian pilots out. With the Junkers in tow, Majors Dave Houston and Brian Cunniff led it back to Winnipeg. Magneto problems grounded the vintage aircraft for a short time in Minneapolis, Minnesota, but thirty-one days after departing Orlando, the Junkers finally made it back to Canada, where it has became one of WCAM's biggest attractions.

On Saturday, August 20 1983, the Squadron celebrated its 50th anniversary with various activities including a static display. A considerable effort had been made to secure for display all twelve aircraft that had seen service with 402 Squadron since the beginning in 1932. However, while still more than a respectable turnout, only seven aircraft types were on site when the anniversary celebrations began. They included a de Havilland DH60 Gipsy Moth, a DH82 Tiger Moth, a North American Harvard Mk II, a North American P-51D Mustang IV, a Beechcraft Expeditor (C45), a Canadair T-33 Silver Star, and of course the Douglas Dakota DC-3.

THE PRIZE JUNKERS JU52/3M IS RETURNED TO WINNIPEG THANKS TO A 402 CREW. THE AIRCRAFT CAN BE SEEN IN ITS CANADIAN MARKINGS AT THE WESTERN CANADA AVIATION MUESUM IN WINNIPEG

The de Havilland DH60 Gipsy Moth had served the Squadron in its early days. Mr. Watt Martin of Milton, Ontario was the owner and operator of the Moth on display for the Squadron's 50th Anniversary celebrations, which arrived in Winnipeg courtesy of a 429 Squadron C-130 Hercules transport aircraft.

was then owned and operated by Mr. and Mrs. Bill Bailey of Calgary, Alberta.

In 1984, one of the busiest elements of 402 Squadron was its Band. The Canadian Air Force was celebrating its 60th Anniversary with a number of activities, and the 402

DE HAVILLAND GIPSY MOTH

Perhaps one of the most exciting aircraft to see service in the Second World War and more recently, with 402 Squadron during the postwar period leading up to 1956, was the North American Mustang IV. The aircraft displayed during the Squadron's reunion served with the Bolivian Air Force and

Squadron Pipe and Drum Band performed at many of them. As is typical for such an important, sought after military capability, the Squadron's band performed at a number of activities off base as well. All in all, the Band racked up nearly three formal engagements per month, beginning with a Change of Command ceremony and concluding with Remembrance Day ceremonies at the St. James Legion. LCol M.S. Joyce, CD, Commanding Officer 402 Squadron, took his retirement, making room for LCol L.E. Olson, CD. Finally, for the first time in the Squadron's history a contingent of "First-Aiders" entered as a team in the Regional First Aid Competition for Regular and Reserve Force personnel. Remarkably, the Squadron's team placed a close second to the Princess Patricia's Canadian Light Infantry team from the neighbouring "South-side" base in Winnipeg.

The strong finish for the 402 squad permitted them to compete for the Mary Otter Trophy. The 402 team consisted of Cpls R. Smythe, M. Fry, K. McKinstry (team Captain), and P. Lockhart, and Privates (Ptes) T. Olson and H. Leask.

As the end of the Canadian Navy's 75th anniversary celebrations drew nigh in 1986, the Air Reserve and specifically, 402 "City of Winnipeg" Squadron was well represented at the closing ceremony. As the *Sentinel* reported in their 1986/3 issue "…it was a belated birthday gift…" with which the 402 Squadron Pipe and Drum Band surprised the navy celebrants. The previous November, the ship's company of *HMCS Chippawa* had just lined up for divisions parade, when out of nowhere the pipes and drums of the 402 "City of Winnipeg" Squadron struck up, taking the parade completely by surprise. "The airmen were on hand to present the sailors with a 75th anniversary gift: a belaying pin fashioned after those on *HMS Victory* made from heartwood of Canadian oak." The 402 Band had actually been smuggled in the back door unbeknownst to all, and the pipers had even managed to tune up entirely undetected. All in all this was one covert operation to be proud of.[6]

As the last NATO member to operate the Dakota aircraft, Canada went into the record books doing precisely that. Winds were from the northwest at 20 miles per hour, and the temperature was -28°C – a cold winter's day in Winnipeg. Nevertheless, LCol L.E. Olson, Capt C.V. MacIntyre and Cpl G. Lloyd manhandled the airplane to a stop; fifty-years-to-the-second after the first Dakota had taken to the air in California, on 17 December 1935.[7]

IT'S A LONG WAY TO THE COCKPIT. A LOOK AT JUST HOW HIGH THE FLIGHT COMPARTMENT OF THE DAK SAT OFF THE GROUND

In April 1986, final preparations were underway to close down all the remaining Air Reserve messing facilities at Canadian Forces Base Winnipeg. True to the increased effort to integrate Regular and Reserve Force personnel, it had been decided, in an effort to save money, that Reserve Force personnel should and would partake of Regular Force messing facilities. Consequently, memorabilia belonging to the 402 Air Reserve was recovered from those 'about to be closed' messes, for preservation in the Western Canada Aviation Museum. Sgt G. Jones, Capt F. Woodward, Honourary Colonel H.N. Scott, Chief Warrant Officer

(CWO) G.T. Wilson, Maj R. Bodnarek and Master Warrant Officer (MWO) D. Weber, played a strong part in these efforts to hand over artefacts to the president of the museum, Mr. John Davidson. In 2005, the items from the old mess, including the photographic collection, were repatriated back to the Squadron's base of operations at 16 Hangar.

The closure of "Building 6" – that building generally referred to as the home of 402 Squadron – did not pass without comment. Fred Cleverley, in a *Winnipeg Free Press* article lamented the closure, making special note that "Building 6" was just about the only wartime facility that continued to house its original tenant. In essence, Cleverley spoke for many who had developed an attachment to the special facilities the building provided to members of 402 Squadron. The building's "Spitfire Lounge" was full of memorabilia about which Cleverley and others were genuinely concerned. Perhaps the building's allure resulted from its link to "the largest air training operation of all time – the British Commonwealth Air Training Plan." As has been known to happen, more often than not "all good things come to pass."[8]

402 RESERVE MESS, HOME OF THE INFAMOUS "SPITFIRE LOUNGE"

From an operational perspective, a certain exercise took place in Canada's far north in the early summer of 1986, highlighting the true meaning of the term 'airpower'; in the sense that 'airpower' is about the synergy realized from combining air assets focused on a common purpose. Since early 1985, the Cold War adversaries best represented by the Soviet Union and the United States of America had been deeply involved in nuclear arms talks. The Soviet's Andrei Gromyko and America's Secretary Schultz met often to discuss how best to adjust stocks of conventional and other weapons. Cruise missiles were but one of the items up for discussion. The CFB Winnipeg newspaper, the *Voxair*, was clear in its claim that "…the revival of low-level attack technology…" had left our arctic flank vulnerable again. So the airpower assets of 425 Squadron Bagotville and 402 Squadron Winnipeg were pressed into action as part of an effort to test Air Command's ability to eliminate this perceived defensive weakness.[9] 402 Squadron's Dakotas deployed in support of Canada's fighter aircraft team, thereby filling in an important logistics capability for which Hercules was deemed less efficient.

Besides these important activities, the Squadron still had time for important social events and activities. 402's third annual "Golf/Picnic/Baseball Day" was quite well organized, thanks to CWO G.T. Wilson and Cpl D.M. Wilson. Cpl Dave Wilson was apparently and inadvertently rewarded for his great efforts when he placed first in the Squadron golf tournament. The annual Riverborne Raft Race also saw Squadron participation. The 402 team very nearly placed first in this gruelling competition, holding onto second place for more than two-thirds of the lengthy race. Unfortunately, muscle power waned and the team failed to show, coming in fourth place, still quite respectable out of 26 teams.

On 23 November 1986, 402 Squadron held a special ceremony marking the 25th anniversary of the day its battle honours, or "colours" were received. Former 2402 member, Brigadier-General E. I. Patrick, at the time serving as Chief of Staff, Operations, at Air Command Headquarters, was delighted to serve as Reviewing Officer for this auspicious occasion. Remarkably, the Standard Party consisted of WOs Enright, Scott and Sterry; three NCOs who had served with the Winnipeg Bears twenty-five years earlier when the Squadron first received its colours.

A number of interesting events took place in 1987. For one thing, the Squadron paraded in "blues" – a new air force uniform reflecting the government's support of the need to allow the CF to revert to environment or service dress uniforms again. Just as an air force squadron's battle honours serve as a symbol of something worth beholding, distinctive service dress uniforms to many are considered just as symbolic and important.

In a superb gesture, 402's Honourary Colonel H.N. Scott, DFC, CD, presented the Squadron with $1,500.00 on behalf of the Canadian Fighter Pilots' Association. This important and timely donation went a long way toward the purchase of new band uniforms. Scott was a decorated bomber pilot and

HONOURARY COLONEL NEIL SCOTT, DFC

former member of 402 Squadron. He had flown with 61 Squadron (RAF) during the war, and wrapped up his wartime experience flying the de Havilland Mosquito with the Pathfinders.

The third in a series of land force operational exercises dubbed "Rendezvous" was underway in Wainwright, Suffield and Cold Lake, Alberta. While somewhat smaller than Rendezvous '85, Rendezvous '87 still had over 12,000

Canadian Forces personnel taking part. An officer from 402 Squadron deployed as part of the public information and public affairs mission.

In an initiative fronted by the Squadron's Honourary Lieutenant-Colonel S. J. Chapman, CD, a 402 delegation travelled to the ancient St. Clement Danes Church in the City of Westminster, London, UK, to present a Squadron Crest floor plaque, and to see to the consecration of the Squadron Standard. The church had been virtually destroyed by enemy bombs in 1941, however, after the war, the Royal Air Force made a public appeal for funds. By 1958, the church had been completely restored, owing to the efforts of the Royal Air Force and all those who had supported the cause. From that moment on, the St. Clement Danes Church has served as the Central Church for the Royal Air Force. Evidence of tokens from all Common-wealth air forces that participated in the Second World War, can be found throughout the church.

HONOURARY LIEUTENANT-COLONEL S.J. CHAPMAN

Troop and light cargo transport, VIP transport and support to other Canadian Forces' tasking continued to describe the role of 402 through 1988. However, the Squadron was preparing itself for changes to its role as a result of the imminent retirement of the Dakota in favour of the new de Havilland Dash-8. Such a change was expected to bring with it the requirement to provide training support to the Canadian Forces Air Navigation School (CFANS) and for the Squadron to assume responsibility for training all Dash-8 pilots. Commensurate with these activities, total flying hours for the Winnipeg Bears fell from traditional averages of some 4,000 hours per year to 1,698.2 hours in 1988. The primary reason for this decline had everything to do with 402's aircraft strength falling from a high of nine aircraft to four as the year unfolded.

Technical training of ground support personnel was a significant challenge to which 402 responded admirably. Training for the 1930s/40s vintage Dakota had to be raised to a state-of-the-art '90s level, in anticipation of the arrival of the Dash-8. A committee responsible for setting up the Dash-8 Master Implementation Plan (MIP) was formed. Meeting throughout the year, the MIP committee, consisting of personnel representing every facet of 402 and its roles, effectively provided a path for the Squadron to follow in their efforts to field a new aircraft and remodel their approach to new operations.

On 10 March, 402's annual mess dinner was held. Guest speaker for this felicitous event was the Commander Air Command, Lieutenant-General (LGen) L.A. Ashley, CD. Shortly thereafter, Officer Cadet (OCdt) D. Collette was promoted to Lieutenant. Collette held the distinction of being the first 402 pilot student to graduate to wings-standard through the Reserve Pilot Training Program in over fifteen years.

On June 26, command of the Squadron changed hands from LCol J.M. Symonds, CD, to LCol R.W. Patrick, CD. LCol Bob Patrick had a long and distinguished air force career as a pilot with both the regular and reserve force. After retirement in 2001, he maintained his connection with the Squadron by serving as President of the 402 Squadron Association. As President, he played key roles in many Association initiatives including this history. In the meantime, the Multinational Force and Observers (MFO), a non-United Nations mission in the Sinai Peninsula, was in place to ensure that exigencies of recent Camp David peace talks between Israel and Palestine continued to be respected. Canada had begun to support the MFO mission, providing a helicopter squadron, air traffic control personnel and other logistics support. Two personnel from the Squadron: Master Corporals (MCpls) R.S. Schwindt and J.S. O'Leary, were assigned to the mission during the summer of 1988.

THE SQUADRON'S OUTGOING SWO, CWO GARY WILSON, PASSES THE PACE STICK WHILE THE NEW SWO, CWO DENNIS WEBER, ACCEPTS CONGRATULATIONS FROM CO LCOL BOB PATRICK

In 1988, Canada's Air Command and Canadian Forces were seriously considering a greater integration of Reserve and Regular Force personnel in a model described as the "Total Force Concept." Fifty years earlier, when Canada declared war on Germany in 1939, the Royal Canadian Air Force consisted of a paltry 2,200 personnel – hardly enough to wage combat operations against such a formidable foe.

Thankfully, the Reserve Force consisted of 966 trained personnel, representing almost half the size of the small Regular Force. Canada's relatively significant Reserve Force therefore was an essential capability. Without this trained core of personnel, it would have been much more difficult to establish initial conditions of strength for the Royal Canadian Air Force in war. All this to say that 402 has been one of the longest serving units within this Reserve Force structure that Canada and its air force have come to depend upon.

As the nine Dakotas soldiered on with 402 and the Instrument Check Pilot School (ICP) the de Havilland Dash-8 waited in the wings. Two Dakotas were touring the country in Second World War markings as part of a farewell ceremony in advance of the official 31 March 1989 retirement date. On 28 September 1988, a Fredricton, New Brunswick newspaper carried a story about the impending retirement of the venerable Dakota.[10] The Dakota that flew through Fredericton had seen previous service with 435 and 436 Squadrons in the Far East. As part of the retirement tour, the

402 OPENS HER DOORS TO THE PUBLIC IN 1988, FOR A FINAL LOOK AT THE SQUADRON'S DAKOTAS PRIOR TO THEIR RETIREMENT

"Gooney Bird" entertained select passengers for a brief ride from Moncton to Fredericton. Onboard during the short flight was the winner of the Sir James Dunn Award of Excellence – an award presented to the top Air Cadet in 333 Lord Beaverbrook Squadron. Stephen Damery, a youngster at the time, fondly remembers his trip on the famed Dakota aircraft.

ONE OF THE TWO DAKOTAS PAINTED IN WARTIME COLOURS AND USED TO CELEBRATE THE AIRCRAFT'S MANY YEARS OF SERVICE WITH THE RCAF / CF

Stephen recalls:

[that time] seemed to pass quickly and we made our approach at the Fredericton airport before we knew it. 'Why did you wake me up, you old #$*%^!, the Flight Engineer protested to his fellow crewmember. However, he decided that it was best to get up. I decided that it was best to keep my laughter to myself. The crew prepared the plane for the large number of spectators and veterans that came [to see the airplane]. Like Moncton, we were greeted with the same warm welcome and high level of enthusiastic crowd.

A BEAUTIFUL AFTERNOON ON THE PRAIRIES FOR FLYING. THE CC AND CT-142S IN THEIR ALL OVER GREY FINISH

The 402 Squadron crew left a tremendous impression on him. To this day, Stephen still cherishes the photos and mementos, including a certificate signed by the Aircraft Commander, Capt R. Hebert, signifying that on 15 March 1989, Stephen Damery had indeed "been privileged to fly in a 402 Squadron Time Machine." A total of twenty-two deserving cadets were given flights on the Dakotas during

CHAPTER 4

THE SQUADRON AT THE DAKOTA CLOSE-OUT IN 1989

THE SQUADRON'S "THE TIME MACHINE" DAKOTA CREST

their farewell tour. The Dakota began service with the Squadron in 1975. Throughout its life with 402, it served as an aircrew training aircraft, light transport duties, search-and-rescue, and as a drop aircraft for the Canadian Forces Parachute Demonstration Team, the "Skyhawks." A life extension to 1998 was in effect when the aircraft was retired from service in 1989.

In October 1988, the Squadron held its annual Jack Reeve Memorial Dinner. MCpl D. M. Wilson was selected as the top Air Reservist. In recognition of his efforts, MCpl Wilson was presented with a cheque for $100.00. In a gesture fully reflecting the current focus on the approaching end of one era and the beginning of a new one, Wilson donated the funds to the Western Canada Aviation Museum 402 Squadron Historical Project.

Though the Cold War began to fade as the 1990s unfolded, little in terms of 402's role changed. Operational support to the Canadian Forces Air Navigation School and airlift

184

WE STAND ON GUARD

support to Air Command were the primary missions that made up the Squadron's role in 1991. As 402 entered its 60th year of operations, the Winnipeg Bears justifiably boasted being one of the oldest squadrons in the Canadian air force.[11] The CC and CT-142 Dash-8s were the backbone of the Squadron, but technicians trained on the Dash-8 were necessary for the maintenance of the aircraft. That's where the Technical Training Unit (TTU) of 402, a sub-unit responsible for many in-house courses related to maintenance of the Dash-8, showed its importance. The TTU remains a vital part of the Squadron.

Significant budget constraints led to some important organizational decisions being made in the closing days of 1989 and opening weeks of 1990. One of those decisions led to an increased emphasis on the importance of Dash-8 operations carried out by 402 Squadron. 429 Squadron and their Hercules aircraft based in Winnipeg were deployed permanently to Trenton. Because the Hercules had provided support to the Air Navigation School, the task now fell to 402's CT-142 Dash-8s.

THE EARLY THREE COLOUR WRAPAROUND CAMOUFLAGE SCHEME WORN BY THE TWO CC-142s EARLY IN 402 SERVICE

SHUTTLING THE PRIME MINISTER, BRIAN MULRONEY AND HIS WIFE MILA, 1 MAY 1992

Among the noteworthy missions carried out by the Squadron's crews were flights involving the Prime Minister of Canada, the Right Honourable Brian Mulroney, PC, MP. Mr. Mulroney had business in Gatineau, Charlevoix, Quebec City and Montreal, for which he came to depend on 402 Squadron for their positive responsiveness and professional demeanor. Princess Anne's Royal Tour of Canada also benefited from the presence of the Squadron. Her entourage travelled between St. Anthony, Deer Lake, Shearwater and Trenton, Nova Scotia.

SGT JOE COSME AT THE FLIGHT ENGINEER'S POSITION ON CC-142 TRIP

Much work was being done at 402 to attend to its most important resource – personnel. Four officers and forty non-commissioned officers were recruited by 402's reserve flight. Administration sections were undergoing significant organization change, and the maintenance flight had instituted a new award to honor the talents of its top-performing member. The first winner was Pte M.A. Hoppensack. A total of fifty-three students completed various courses offered by the 402 TTU, with another forty-nine completing servicing training. Finally, two additional events for 1991 also deserve mention. First, the Squadron held an Employers Appreciation Day, during which Capt H. Chase was honoured with a Certificate of Appreciation for the outstanding support that Air Canada had provided to the unit. A great deal of coverage on local television news stations resulted, reflecting the very positive way the Winnipeg community received 402's efforts to honour local employers. Secondly, at the 18th annual Jack Reeves Memorial Dinner, Cpl P.D. McNorgan was the recipient for 1991 of the certificate and $100.00 for being the "Squadron's Airman of the Year".

In addition to looking after their own six Dash-8s, 402 also tended to the Central Flying School's (CFS) six CT-114 Tutor aircraft for both servicing and snags; as well as servicing the school's two King Air aircraft and one Jet Ranger. Add transient aircraft servicing to these tasks, and it made for a very busy Squadron, with technicians qualified on two aircraft.

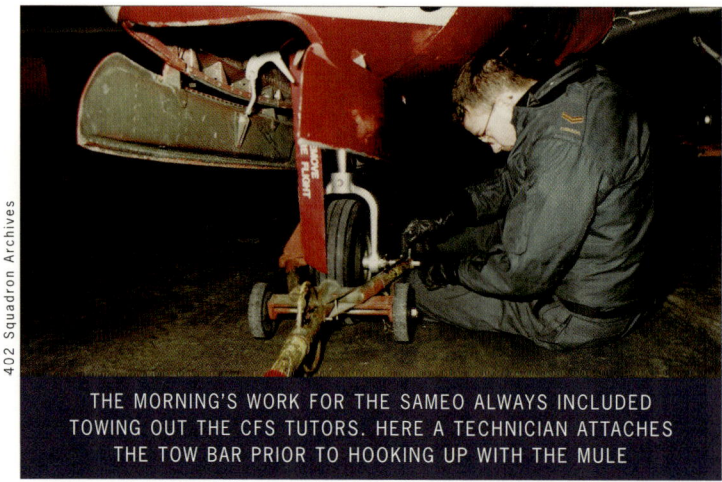

THE MORNING'S WORK FOR THE SAMEO ALWAYS INCLUDED TOWING OUT THE CFS TUTORS. HERE A TECHNICIAN ATTACHES THE TOW BAR PRIOR TO HOOKING UP WITH THE MULE

THIS PICTURE FORMED THE BASIS FOR A PAT MCNORGAN MURAL AT THE EAST END OF THE BREEZEWAY IN HANGAR 16. CPLS DANA CRAIG AND DON THOMPSON PERFORM AN AB CHECK BETWEEN FLIGHTS

402 began these additional tasks after the arrival of the Dash-8s, when the Base Aircraft Engineering Organization (BAMEO) became 402's Squadron Aircraft Maintenance Engineering Organization (SAMEO). Up until this point, many Reserve Squadron technicians worked within BAMEO, gaining experience on the CC/CT-142s. When taking control of this organization, with its many Regular Force members, 402 really was at the forefront when it came to the "Total Force" concept. "Total Force" was an initiative that saw Reservists trained up to their Regular Force counterpart's levels, and working side by side with them.

HONOURARY COLONEL R.G. JAMES, CD

CPL G.J.R. LACOURSIERE, RIGHT, AND CPL H.N. COMPTON, LEFT, SUPERVISING A STUDENT TECHNICIAN FROM TEC VOC HIGH SCHOOL, ON THE FINER POINTS OF THE DASH-8'S MAIN GEAR

CHAPTER 4

NO VACANCY. THIS REMARKABLE PHOTO SHOWS TWO TRANSIENT TUTORS, FIVE CFS TUTORS, ONE SNOWBIRD, A KING AIR, A CF-18, TWO DASH-8s AND ONE JET RANGER

LCOL CHUCK MACINTYRE RECEIVES THE TRADITIONAL SOAK AFTER HIS LAST FLIGHT AS 402'S CO

The highlight of 1992 may well have been the diamond anniversary weekend, celebrated during the period of 14 to 16 August. About five hundred current and former members gathered to raise a glass on the occasion of this auspicious milestone. The Squadron also held a change of command parade as LCol R.W. Patrick, CD, stepped aside, making room for LCol C.V. "Chuck" MacIntyre, CD. LCol MacIntyre, an Air Canada pilot in civilian life, had an eclectic aviation career, which included flying Trackers off of the aircraft carrier *HMCS Bonaventure*. A celebration honouring the investment of new Honourary Colonel also took place. The outgoing Honourary Colonel, R.G. James, bid farewell as Honourary Colonel P. McGonigle, a former Lieutenant Governor of Manitoba, took over. Finally, the recipient of the Jack Reeve Memorial Award for the airwoman/airman of the year for 1992, was Pte K.L. McDonald, who was presented with the award during the annual Memorial Dinner. Anyone of Scottish origin might be justifiably proud of this MacIntyre-McGonigle-McDonald sweep, amongst the Squadron's ranks of 1992.

CT-142 SPORTING THE DOUBLE GREY SCHEME WHILE THE SQUADRON CELEBRATED HER 60TH ANNIVERSARY IN 1992. THE 60TH CREST IS TO THE LEFT OF THE MAIN DOOR

JACK REEVE MEMORIAL AWARD WINNER PTE K.L. MACDONALD

The VIP missions continued in 1996, highlighted by flights to Argentina involving the Chief of the Defence Staff and the Minister of National Defence; and the Commander of the Russian Air Force, Lieutenant-General Krivolapov. Additionally, 402 provided some airlift assistance in support of the Land Forces Command and Canadian Forces deployments to Haiti. The total number of passengers airlifted was 4,537 of which 521 were classified as VIPs.

THE CC-142 FLIGHT ATTENDANTS COURSE.
LEFT TO RIGHT: MCPL DAVE WILSON, CPL JENNIFER CHAMBERS, SGT RICHARD DUBE, MCPLS DIANE CLAYTON AND JEFF O'LEARY

PART OF THE 60TH ANNIVERSARY CROWD AT THE WINNIPEG CONVENTION CENTRE IN 1992.

On 18 January 1997, LCol D.G. Lamb, CD, took command of the Squadron from LCol P.G. Rawlings, CD. The new CO had served with the United Nations in Egypt, and flown Kiowa helicopters with 10 Tactical Air Group (10TAG) as a member of 400 Squadron. At the time he took over 402 Squadron, LCol Lamb was also an Air Canada A-320 pilot with over 15,000 hours flying experience behind him. The strength of 402 at the time of the change of command was 148 Regular Force, 100 Reserve Force and 3 civilian personnel.

BGEN (RET'D) PETER DESMEDT, A WINNIPEG CITY COUNCILLOR, RECEIVES THE OLD 402 STANDARD FROM 402'S COMMANDING OFFICER LCOL DAVID LAMB. THE OLD STANDARD REMAINS ON DISPLAY AT WINNIPEG'S CITY HALL

Members of the newly commissioned multi-role patrol frigate *HMCS Winnipeg* paid a trip to their namesake's city, where "The City of Winnipeg" Squadron and city hall entertained them. 402 later repaid the visit with a Dash-8 flypast and band performance for the crew of the warship.

PART OF THE CREW OF *HMCS WINNIPEG* VISIT WITH 402'S CO LCOL DAVID LAMB

In 1998, sixty-eight members of 402 Squadron pitched in to deliver Christmas hampers to those in need. This was followed by the efforts of Squadron personnel who built a house for flood victims as part of "Habitat for Humanity" following what had been declared the flood of the century. Along with the help of twenty Squadron volunteers, Flood victims André and Suzanne Tetrault had their home replaced.

The introduction of Windows 95 and Microsoft Office enabled the Squadron to formally link with the Wing Area Network, resulting in a significant improvement with the lines of communication. For the first time, 402 Squadron

A DASH-8 CARRYING A MOBILE REPAIR PARTY PAYS A VISIT TO NORTH ISLAND NAVAL AIR STATION IN SAN DIEGO, MARCH 1995

had its own home on the internet, with much thanks to the efforts of computer specialist Cpl S. Roberts. Cpl Scott "Big Daddy" Roberts' development of an internet site did not go unnoticed, and he received a Commanding Officer's Commendation for his service to 402. The VIP flights continued with Her Royal Highness Princess Anne, Chief of Defence Staff General M. Baril, and a number of other General Officers all benefited from 402's professional service.

Other Squadron members continued to be recognized for their achievements and generosity. Cpl M.G. Hamilton, a 402 Avionics Technician (AVS) and aviation history buff, discovered a vintage Bristol Bolingbroke aircraft not far from the Winnipeg bedroom community of Headingley. He approached the farmer-owner of the Second World War vintage medium bomber and trainer, who agreed to sell it for $100.00 rather than scrap it. Cpl Hamilton then magnanimously offered it to 17 Wing Heritage, who accepted the gift and had the "Boly" transported to a field on the base with plans for a restoration someday. As

CPL MARCEL HAMILTON, WHOSE EFFORTS SAVED THE VINTAGE BOLINGBROKE, STANDS BY THE VERTICAL STABILIZER OF THE SECOND WORLD WAR BOMBER

an acknowledgement of his efforts to preserve a piece of RCAF history through his gift to the Air Force, Cpl Hamilton was awarded a Wing Commander's Commendation.

THE BRISTOL BOLINGBROKE AT 17 WING, AWAITING RESTORATION

Activities within the Maintenance Flight continued at traditionally significant levels. The AF9000+ initiative was well underway and it appeared that the Squadron was leading the Air Force in so far as establishing this industry standard quality management undertaking. In 1998, the Squadron was poised to receive formal registration. "Operation Phoenix" also kept the Maintenance Flight busy. The aim of this Operation was to re-engineer aircraft support processes in an undertaking somewhat akin to the AF9000+ initiative. 402's SAMEO were fighting against contracting out the maintenance work for the Dash-8 fleet. The Squadron proved to Air Command that its maintenance organization was more cost effective and a better alternative. Finally, "Operation Determination" was of particular interest to three 402 Squadron maintenance personnel. Iraq's discredited leader, Saddam Hussein, continued his efforts to scuttle or thwart UN weapons' inspectors, in the wake of his failed attempt to overtake Kuwait. As part of the international effort to respond to Hussein's posturing, additional "Tanker" aircraft were needed in order to keep fighter aircraft in the air. To that end, 402 provided three maintenance personnel to support the operations in the Middle East.

For the first time, 17 Wing participated in the annual Nijmegen March in 1998. The Wing's team of eleven participants included nine Squadron members – more than a respectable contingent, to be sure. The team included Second Lieutenant (2Lt) T. Neal, WO L.J.J. Viens, MCpl D. Murray, Cpls J.A. Bell, P. Charbonneau, K.S. Hubbard and P.J. Van Helvert, and Ptes A.W.N. Semper and E. Penner.

Training began in mid-April with the aim of trekking as many miles possible, to tone up for the demanding Nijmegen March. The team averaged approximately 80 kilometres per week rain or shine, finishing their training mid-July. Some six thousand military personnel from twenty nations participated in the march that year, and the Squadron's contingent, which made up most of 17 Wing's Team, successfully completed the gruelling four-day test.

The new millennium ushered in significant changes for the Winnipeg Bears, primarily in terms of fleet reductions and consequent changes to the Squadron's roles. As a result of Defence Planning Guidance 2000, the two passenger models of the Dash-8 were declared surplus, withdrawn from service in 2000, and eventually sold to the private sector in 2002. Thus ended the medium transport and VIP executive transport roles. The four remaining CT models soldiered on, and with a mandated yearly flying rate (YFR) of 2875 hours, continued to fly and support the CFANS training syllabus which now includes Airborne Electronic Sensor Operator (AESOp) training and CT-142 pilot conversion, proficiency and upgrade training.

Major-General (MGen) (ret) R. Linden, CMM, CD, became the Squadron's new Honourary Colonel in 2004, taking over from Pearl McGonigle. He was an obvious choice, having previously served with 402 in 1972, flying Otters and later Dakotas as Flight Leader. On his promotion to the rank of MGen, Linden was appointed Chief of Reserves, an appointment that dovetailed nicely with the "Total Force" nature of 402.

CT-142 OVER THE ROCKIES WITH THE NEW DARK BLUE SCHEME

In 2006, the Squadron operated with a personnel strength hovering in the 185 to 210 range comprising Regular Force, Reserve Force and two civilians working together in every aspect of the operation. The Squadron Pipe and Drum Band maintained their very busy schedule in a wide range of venues in support of the Squadron, 17 Wing and of course, the City of Winnipeg. The Band continues to participate in numerous concerts, parades, mess dinners and many other public performances. These events cannot be accomplished without the participation of the Band's civilian volunteers who make up half of its strength.

As a tribute to the 60th Anniversary of D-Day in 2004, and to acknowledge the part that 402 played in the Normandy Invasion, the Work Shop technicians applied vintage black and white "invasion stripes" to Dash-8 805. The stripes were applied in the two official positions, the aft fuselage ahead of the empennage, and outboard of the engines on both wings, harkening back to that pivotal day in history when the Squadron's Spitfires donned similar war paint and kept the skies above the beaches, clear of enemy aircraft. All allied aircraft involved in the invasion were to be sporting the stripes for ease of sea, ground and air identification. The plan was for Dash-8 805, resplendent in her new "old" paint scheme, to participate in the flypast over Winnipeg to mark the significant anniversary. Unfortunately, a technical problem kept 805 on the ground that day, thus preventing the public from seeing the Squadron's toast to one of her cherished battle honours.

402 Squadron personnel received a host of honours and awards in recent years. In addition to the Jack Reeve and

CT-142 805 PAINTED WITH INVASION STRIPES IN 2004

Vernon Bastable Awards, Squadron members received: Canadian Forces Decorations (CD) CD1 and CD2, Canadian Peacekeeping Service Medals (CPSM), Queen's Golden Jubilee Medals (QGJM), Canadian Forces Suggestion Awards, Prime Minister's Certificates and Certificates of Service to retiring members. Flight Safety Good Show Awards, Flight Safety For Professionalism Awards, a (Chief of Defence Staff) CDS Commendation, Commanding Officers Commendations, and the Squadron Aircraft Maintenance and Engineering Organization's (SAMEO) Awards round out a very impressive list.

The "City of Winnipeg" Squadron's ties with the community remain firm. In 2004, Bruce Middle School, in the Winnipeg suburb of St. James not far from 17 Wing, formed an association with 402. The idea was the brainchild of principal Bill Cann, who thought that being a part of the military community, an association would be beneficial to both school and squadron. The school was then divided into four "houses" with each being named after fighter aircraft that 402 had operated: the Mustangs, the Spitfires, the Hurricanes and the Vampires. With many Squadron members continuing to help out with Bruce's annual science fair, and an added historical Air Force display in the school's atrium, the relationship between Squadron and school has only become stronger.

SGT PAT MCNORGAN, SECOND FROM RIGHT, PROVIDED THE ART FOR THE FOUR "HOUSE" POSTERS AT BRUCE MIDDLE SCHOOL

The aircraft livery changed with the CT-142s repainted from light grey to high gloss training blue. This was done to increase visibility since the aircraft are flying in a relatively busy, low-level air structure. Squadron members also changed their look as the aircrew transitioned to olive green flying clothing, while maintenance and support members were kitted with camouflage "CADPAT" (Canadian Disruptive Pattern) gear. The blue beret, dark blue tee shirt, nametag and rank, are worn with the CADPAT to distinguish Air Force members, but most civilians, with the preponderance of green in the uniform, see only an Army soldier. Only the service dress uniforms, Air Force blue,

FORMER 402 SWOS. LEFT TO RIGHT: CWOS DAVE MCALLISTER, MIKE WEBB, TOM "BULL" SECRETAN, AND DENNIS WEBER

Army green or Navy blue, remain to readily identify the three different service affiliations of Squadron members.

402 has taken up an important role in the defence of Canada operations. 1 Canadian Air Division/Canadian NORAD Region Headquarters identified Winnipeg as a Forward Operating Base for CF-18s in the NORAD Air Sovereignty Alert mission. An agreement between 4 Wing Cold Lake and 17 Wing Winnipeg, contained an inclusion that 402 Squadron SAMEO members provide servicing for the fighters whenever the CF-18s were in the Manitoba Capitol.

Maintenance personnel have been trained and exercised in this very important duty through Air Sovereignty Alert (ASA) aircraft deploying regularly to Winnipeg.

The success of the Squadron depends on each section; however, one division, the SAMEO, continued to carry out a tremendous amount of work. A constantly evolving organization, they are always striving for new ways to achieve perfection. All of this effort is necessary to keep up with the changing taskings of the Squadron, and the aging of the aircraft. Various internal sections have been formed, realigned, amalgamated and even eliminated. Whether it be aircraft repair, inspection, modification or painting; whether it be logistical support or airworthiness issues; whether it be maintenance or personnel support to other Wing units including 435 Squadron, 440 Squadron Yellowknife, 3 Canadian Forces Flying Training School, Canadian Forces School of Survival and Aeromedical Training, Central Flying School, Wing Operations or 1 Canadian Air Division/Canadian NORAD Region Headquarters; whether it be infrastructure improvements; the SAMEO organization responds to all challenges in the true Winnipeg Bears 'can do' tradition. Logically, the

MCPLS TIM SORENSEN AND JIM HENNESSEY AT WORK IN THE "BLISTER"

training of Squadron members fall within these mandates, and takes place in a variety of areas including: Fall Protection, transportation of dangerous goods, Air Maintenance Policy (AMP) training, CT-142 Maintenance Manager's course, CT-142 Line Servicing course, CT-142 Aviation Systems (AVN) On-Type Maintenance course, CT-142 Avionic System (AVS) On-Type Maintenance course, Confined Spaces Entry course, De-icing/Bucket training, Compliance Awareness training, Conflict Resolution, Workplace Hazardous Material Information System (WHMIS) training, WASF training, Mobile Support Equipment (MSE) Safety course, Driver Instructor Examiner course, Flight Safety course and Human Performance in Military Aviation (HPMA) training and Squadron Indoctrination Course, to name but a few.

THE "GONZO" CREST, WHICH WAS WORN ON FLIGHT JACKETS AND FLYING SUITS

While the aircraft exterior remains relatively untouched, except for the new paint scheme, the internal equipment and systems constantly change. Upgrades, modifications and new equipment add to the already impressive capability of 402's Dash-8s. The cockpit instrumentation, navigation and instrument approach equipment systems in the CT-142 has always been ahead of most Canadian military aircraft with the Flight Management System (FMS), Global positioning System (GPS), Category II precision instrument approach capability and Terrain and Collision Avoidance System (TCAS); the continual improvements only make it better. Even more upgrades are planned, including latest generation FMS and GPS, a Terrain Avoidance Warning System (TAWS), installation of a "pulselite" collision avoidance system, and replacing the Flight Data Recorder/Cockpit Voice Recorder (FDR/CVR).

LONG-TIME 402 MEMBER SGT OLLE FRITSCH IN FRONT OF CT-142 804

Because the Dash-8s are flown on routine missions, the maintenance support to keep them flying has been a remarkable undertaking. These aircraft are aging gracefully, but at the same time, the person-hours required to keep them in the air will increase as the years roll on, throwing down even more challenges to the maintainers. Along with the daily inspections, the Squadron maintenance men and women are kept busy with airframe and engine periodic and supplementary inspections, major component changes that include engines and landing gear assemblies, while constantly dealing with many minor snags that pop up on virtually every flight. The current approved Estimated Life Expectancy (ELE) for the CT-142 had, at the end of 2006, been approved by the Chief of Air Staff for a ten-year increase, moving the original life-span of 2011 ahead to 2021. Also in 2007/2008, the CT-142 will be given a half-life inspection, based on the Transport Canada (TC) Dash-8 half-life depot level inspection. This half-life inspection is equivalent to the civilian "D" check that is necessary at the 40,000-flight hour mark. Although the Dash-8s are far from the 40,000 flight hours now being experienced by civilian operators, 402 operates the aircraft in a more robust manner than the civilian airlines (on average 11,000 flight hour per airframe). Along with the introduction of the CFANS Basic Air Navigation Course (BANC) training syllabus, that includes mission profiles requiring flight at 1000 feet AGL, it was wisely thought to look into whether there were any gremlins at work on the structural fatigue life of the aircraft. CT-142 805 was provided with an accelerometer; the data was collected over the year to try and find information on the increased turbulence/manoeuvring effects found at lower altitudes. One of the first reports from Bombardier pointed out that the fatigue damage per flight under Air Force operation is about twice as high as the damage per flight during a typical airline commuter trip. This is the rationale for the twenty-year mid-life inspection requirement. This program will mean less flying for each aircraft, bringing with it some very real scheduling challenges for the Squadron.

Commanding Officers come and go on a regular basis, as is military policy. 402 Squadron is no exception, other than the fact that a Reservist has historically led the Squadron.[12] LCol D.G. Lamb, CD, turned the Squadron over to LCol B.M. Doyle, CD, in 2000. LCol S.L. Schock, CD, then assumed command in 2003, becoming the first Regular Force Commanding Officer in some years. In 2006, he relinquished command to LCol R.T. Witherden, M.B.,CD. LCol Witherden was decorated with the Medal of Bravery for piloting the 1984 rescue of eleven sailors from a sinking vessel, the MV *Ho Ming 5* in the middle of a hurricane off the grand banks of Newfoundland. Once again, 402 had a reservist in command of the Squadron, although LCol Rick Witherden had many years of regular force experience under his belt.

As part of the celebration of 402's 75th anniversary celebrations, CT-142 803 was given a special paint scheme reminiscent of the Squadron's Mustangs. Both propeller spinners sported alternating vertical circles of the 402 colours, blue and yellow.[13] The coloured stripes were also applied to the rudder, and "City of Winnipeg Sqdn" in yellow script, adorned the outboard engine cowlings.

To conclude this story on the life and times of 402 "City of Winnipeg" Squadron is to acknowledge that her history continues to be written. Life in the Squadron goes on, and the members that mark 402's 75th Anniversary continue to take up the challenges that face them. The Squadron's airmen and airwomen of 2007 live and work in an Air Force that S/L Sully and his reservists could have only dreamed about in 1932. Of course, it's taken seventy-five years to get from point "A" in 1932 to point "B" in 2007, and changes don't often happen overnight. Were those airmen of the '30s able to magically look through the lens of time and see their Squadron in 2007, they would be both astonished and proud of what the unit has accomplished over the years. Fortunately, a time machine is not needed for today's Squadron members to gaze back over the road the Winnipeg Bears have taken since that simple beginning. When they do so, they see a Squadron that has continually served not only her country with pride and honour, in wartime and in peace, but also uniquely, her city. There can be no higher accolade for a Canadian Forces Squadron, for the forefathers have set the standard high. From the Sullys and Klaponskis, to the Morrows and Northcotts, to the Reids and MacIntyres, the torch is passed and accepted with a deep sense of duty, gratitude and humility. To march with such company demands it.

"We Stand On Guard"

NOTES

1. John F. Kennedy speech September 12, 1961.
2. Please see page 167 for a complete table of RCAF ranks and their CF equivalents.
3. Canadian Defence Priorities: A Question of Relevance 1972.
4. Yet another nick name fondly applied to the Dakota.
5. Interview with Sgt McNorgan, March 2007.
6. *Voxair*, 5 March 1986.
7. *Voxair*, 19 March 1986.
8. *Winnipeg Free Press,* Monday, April 7th 1986.
9. *Voxair,* 23 July 1986.
10. *The Daily Gleaner* – Fredericton.
11. Both 402 and 400 Squadrons stood up on 5 October 1932. S. Kostenuk and J. Griffin. *RCAF Squadrons and Aircraft.* Pages 80-84.
12. During the Second World War, commanding officers were "Regulars."
13. 402's colours were a tribute to the Winnipeg Blue Bombers.

The band stepping out on the 17 Wing tarmac

402 Association Archives

The 402 Squadron Pipe and Drum Band: The Capture of Hearts

Master Corporal Josephine Sallis

The only drum major the 402 Squadron Pipe and Drum Band has ever had to date, Cliff Cooke, originally joined the Winnipeg Bears in 1967 with a somewhat unusual pairing of trades: tenor drummer/disciplinarian. He's been the Band's Drum Major since 1974. Asked why he stays so long, he answers, with a shake of the head and a smile, "I love a good pipe band." Here then, with the Drum Major's introduction said, here is the story of a very good pipe band.

When John Reay Sr was asked to form the new 402 Squadron Pipe and Drum Band, he looked no further than his earlier creations for its first members, son John Reay Jr, and the Thistle Pipe Band of Winnipeg. John Sr was Pipe Major of the Thistle Pipe Band, a private, self-funded ensemble. Born in Winnipeg, of Scottish immigrant parents, John Sr first picked up the pipes at the age of twelve, and soon after began accumulating trophies. Into adulthood, he

PIPE AND DRUM BAND

continued being a top-notch player, serving twenty-one years in the City of Winnipeg Police Pipe Band, before forming the Thistle Pipe Band. His renown led to his selection as the first Band Master of the yet-to-be-formed 402 Squadron Pipe and Drum Band.

After efforts to start a brass band failed, F/L Carl Friberg, Assistant Director of Bands from Ottawa, recruited John Reay Sr to form the newly authorized 402 Squadron Pipe and Drum Band in the spring of 1954. After months of paperwork, Reay was finally commissioned as a Flying Officer, Band Master for 402 Squadron. Recruiting began in October of 1954, before Reay Sr's commissioning. John Jr, then in his early teens, became the first member of the band, sworn in on 16 December 1954.

F/O JOHN REAY SR, CENTRE, RECEIVING HIS CANADIAN FORCES DECORATION, AS JOHN REAY JR LOOKS ON

While most of the musicians from the Thistle Pipe Band were recruited into 402 Squadron to form the nucleus of the new band, several other young players were also enrolled. These new pipers and drummers would provide the basis for the Band's future achievements. Senior NCOs were appointed and Ex-Pipe Major John Frederick of the 7th Battalion Cameron's (Scottish Rifles), with the new rank of FSgt, became the Assistant Band Master. As commanding officer, W/C M. Reid had the responsibility of appointing a qualified bandsman/bandswoman to the position of Band President (a position that was later permanently changed to Band Officer). F/L R. Campbell was the first Squadron member to hold this position. Many felt it was Reid's persistence that ensured 402 had its own Pipe Band. To honour the Wing Commander's dedicated efforts, F/O Reay chose the *Atholl Highlanders March* as the Squadron March Past. Reid's family was a branch of the Robertson Clan, and the *Atholl Highlanders March* is the Robertson Clan March Past.[1]

SQUADRON CO, W/C MATT REID, IN THE COCKPIT OF HIS VAMPIRE. REID'S LEADERSHIP WAS INSTRUMENTAL IN GETTING THE BAND STARTED

In the beginning, the musicians supplied their own pipes and drums until new instruments could be requisitioned. These original drums were the regular service type, not very suitable for a pipe band, but the proper tension drums were eventually purchased from the United Kingdom. After much consideration, F/O Reay chose the Robertson Bagpipe, which the Band has continued to play over the years. The present day Warnock Chanters (a pipe reed) purchased in 1979, are also still in use. Highland dress was not issued until 1957. Prior to that, the Band paraded and performed in regulation RCAF tunics, No. 5 (dress) 5a (battle dress) and

THE BAND, IN NO. 6 SUMMER DRESS, PLAYING AT THE COLD LAKE, ALBERTA TRAIN STATION IN THE 1950s. F/O REAY SR IS AT THE FAR LEFT

IN PRE-KILT DAYS. THE PIPERS LEFT TO RIGHT: AC2 GEORGE DOUGLAS, F/O JOHN REAY SR, AC2 GRANT, CPL SANDY CHERRIE, SGT MURDO MCLEOD, AC2 JOHN REAY JR, FSGT JOHN FREDRICK

THE SQUADRON COLOURS, POSED WITH THE BAND'S INSTRUMENTS

No. 6 (summer khaki). The original purchase was made possible by a fund created by the Officers of 17 Wing, under Commanding Officer, G/C H.N. Scott, DFC. The original issue was comprised of twenty regulation RCAF kilts, sporrans, hose top, and white spats. While not the initial issue, the Band still sports the RCAF tartan for

EVEN A HALLWAY PRACTICE DRAWS AN AUDIENCE.
A PAIR OF BANDSMEN TUNE UP PRIOR TO A PERFORMANCE

parades and performances, and wore RCAF band tunics well into the 1990s.

Even before the arrival of their new uniforms, the freshly formed band was on the road winning competitions. In May 1955, the 402 Squadron Pipe and Drum Band placed first in the Western Canada Pipe Band Championship in Saskatoon.

1955 AND 1956 WESTERN CANADIAN CHAMPIONS. F/O REAY SR IS CONGRATULATED BY THE SQUADRON CO, W/C GRAY

In addition to the championship, individual members, including F/O Reay and FSgt Frederick, picked up sixteen major prizes. On 1 July, the Band traveled to Edmonton where it won the Western Canadian Championship, then repeated the feat the following year in Saskatoon.

On 26 February 1956, still without their ceremonial dress, the Band performed an Air Force "Retreat" at the Squadron's Change of Command Parade. "For this event No. 400 Squadron Toronto, loaned our band sufficient completed RCAF Highland Dress, which were brought to Winnipeg and returned to Toronto by our own Squadron Pilots and aircraft."[2] According to former Pipe Major John Barbour, the same uniforms were borrowed for the Band's 22 June performance at the RCAF Tattoo in Toronto. The

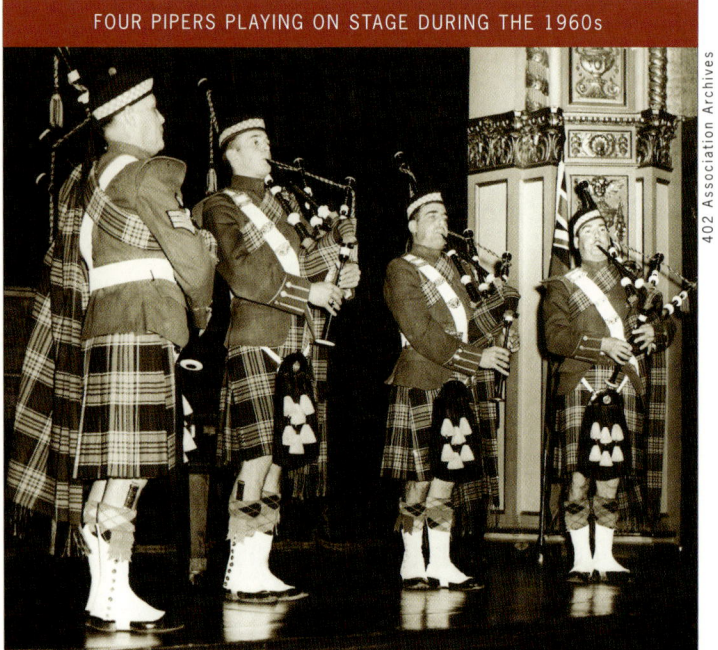

FOUR PIPERS PLAYING ON STAGE DURING THE 1960s

THE BAND GATHERS FOR A GROUP SHOT IN FRONT OF ONE OF THE SQUADRON'S OTTERS

tattoo marked the first time all the RCAF Pipe Bands participated en masse. The following day, the 402 Squadron Pipes and Drums, still utilizing the borrowed uniforms, competed at the Highland Games.

When F/O Reay eventually ran out of age extensions and retired, son John Reay Jr carried on in his father's footsteps. Already commissioned and holding the position of Security Officer with 402, Reay Jr returned to his first love in 1963 as the Band Officer and Band Master. Under his tutelage, the Pipes and Drums began to attend more international events.

Prior to 1960s, travel was a problem because, as a fighter squadron, 402 was not equipped with transport aircraft. The Band had to look elsewhere, and usually hitched a ride with a Regular Component transport aircraft. When the Beech Expediters arrived (and later the de Havilland Otters, Douglas Dakotas and the de Havilland CC-142 Dash-8s), the Squadron could organize its own flights for the musicians. Initially, because of security and transport issues, 402 aircraft could only fly into the closest US base, where a US National Guard aircraft tasked with transporting the Band to their American destination usually met them. While Reay Jr, then a Captain, has many fond memories of warm receptions and welcoming assistance from our southern neighbours, one of the Band flights resulted in some unusual events:

> The band was flying into Minot, North Dakota on four or five Otters in late fall. The Otters were equipped with snow skis outfitted over the wheels. We were just about to touch down when our aircraft suddenly lurched upwards. The landing aborted, we made a pass of the airport and then landed after a second attempt. Our pilot was very upset and after departing the plane, made a quick trip to the airport tower to give a piece of his mind to the air controllers. As it turned out, the Americans had been practicing aircraft carrier type landings and had steel cables strung out across the runway. If one of the skis on the Otter had caught the cable, the aircraft would have nosed into the runway with dire results. The Americans apologized, stating that they had never seen the Otter, ski-equipped, prior to our arrival…

Heading home days later on the same trip south, and just prior to sunset, the Otters were flying in formation at an altitude of only a few thousand feet. They were bucking a strong head wind, traveling slowly with strobe lights blinking. Through the sunset haze, farmers and folks on the ground observed the slow moving craft with lights, and the local radio stations and Police agencies were inundated with calls of U.F.O.s.[3]

The 402 Squadron Pipe and Drum Band were honoured to play in the opening and closing ceremonies of the 1967 Pan Am Games. They proudly performed for HRH, Duke of Edinburgh, Prince Philip in the pouring rain, wearing (and soaking) their new full dress doublets. The uniforms had just been issued during rehearsal the night before. Fortunately, the new duds survived being drenched.

CORPORALS ALL. A MEMORABLE DAY IN THE STATES AT THE SAN ANTONIO, TEXAS, HEMIS FAIR, 1968

Several of the Pipe and Drum's forays into the United States were a result of their 1968 performance at the Hemis Fair

'68 in San Antonio Texas. The six-month world fair was also part of that city's 250th birthday celebrations, and long-time bandsman Neil Barbour remembers not only their performance as part of the Canadian Pavilion, but also the fun they had afterwards having the run of the park and rides on the 92-acre site.[4] During the Fair, 402 Squadron's Information Officer, Lt Keith Cummings (a CBC TV weatherman in civilian life) along with Capt Reay, presented the Canadian Flag to Nicholas W. Classon, who represented the Scottish Society of Texas. The Band went on to win several major competitions in the coming years under the direction of Pipe Major John Reay Jr.

In 1975, Reay Jr retired from the Band because of his civilian job transfer to Calgary, Alberta. WO George Lawrence, promoted from his position as Band Master (Pipe Major), replaced Capt Reay as Band Officer. Besides his work with the Band, Lawrence also helped develop a cadet pipe band for a Royal Canadian Air Cadet (RCAC) squadron in Winnipeg. Here was a situation beneficial for both the RCAC and the Squadron, for many of these cadets eventually found their way into the 402 Band. Under Pipe Major Lawrence, the 402 Squadron Pipe and Drum Band received an invitation to play in the Rose Bowl Parade in Pasadena as part of a massed Pipe Band from Winnipeg. In 1974, during their second performance, they were given the honour of leading the parade. This was a significant achievement because it marked the first time a band other than the Pasadena College Band had led the famous Rose Bowl Parade. Drum Major, Cliff Cooke remembered having a tear in his eye as he witnessed the many Canadian flags flying in the Parade that day.

Their reputation soon made its way overseas, leading to an invitation to the International Military Music Show on 28 May 1978, at the Köln Sports Hall in that West German city.

Now in high demand, the Band was asked by the Canadian Forces Europe (CFE) to support the CFE Pipes and Drums,

LT KEITH CUMMINGS, LEFT, AND CAPT REAY JR, CENTRE, PRESENT THE CANADIAN FLAG TO NICHOLAS CLASSON OF THE SCOTTISH SOCIETY OF TEXAS IN SAN ANTONIO

402 Association Archives

with a specific request for Band member Don MacLeod. Eight pipers flew to Germany providing support to a Guard of Honour for the French Forces' General A. Laurier, and to perform at an International Music Show organized by the German Army for the Cancer Society in Cologne. Their show was outstanding, as reflected in this letter of appreciation from Colonel J.F.Y. Sorel, Base Commander:

> The behaviour and performance of Pipe Major Lawrence and his seven men was beyond reproach throughout. Their accomplishment in memorizing the extensive program required and turning out so smartly each day reflects a great deal of credit on Pipe Major Lawrence, 402 Squadron, and the whole of the Canadian Forces Reserve…Special mention should be made of Master Corporal D. MacLeod who, disregarding substantial facial injuries received shortly after arrival in CFE, played the bass drum on the Guard of Honour, and his bagpipes at the Cologne performance, in spite of his obvious pain. His dedication was much admired by all who knew of his injuries…The military bearing and general professional behaviour of these Reservists has done much to prove our Pipe Major Macpherson's oft-respected contention that use should be made continually in CFE Pipes and Drums of militia and air reserve pipers and drummers.[6]

After leading the Band for the last time in the February 1979 Opening Ceremonies for the Canada Winter Games in Brandon, WO Lawrence retired because of health problems. He continued promoting young musicians through volunteer work with the Navy Cadet Band at *HMCS Chippawa*[7], and even used his contacts at 402 to obtain assistance and instruction for his young charges.

MWO NEIL BARBOUR PLAYING IN DALLAS, TEXAS, 1995

WO N. Barbour succeeded WO Lawerence. Neil Barbour's piping career began in 1951 with the Cameron Highlanders.

CLIFF COOKE LEADS THE BAND DURING THE NAVY'S 75TH ANNIVERSARY CELEBRATIONS AT *HMCS CHIPPAWA*

He served as Pipe Major of the Cadet Band for three years before enlisting in the Senior Band for another three years. After a one-year hiatus, he re-enlisted in the military with 402 Squadron, where he remained until retirement. An independent businessman, former Pipe Major Barbour always felt fortunate and blessed that his business allowed him to devote considerable time to his passion for the pipes. That kind of devotion helped Band musicians continue their active schedule over the coming decades.

The Squadron received a new addition in 1980 with the creation of positions for Highland Dancers. The original four young women were experienced dancers with previous Band volunteer time. The Squadron's reputation and prestige grew with the addition of the women. The dancers obviously stole more than the hearts of their audience, because all four subsequently married Band members. Two of the dancers were the sisters Madeline and Suzy Engleberg. Madeline married future Pipe Major Don Blaine, and Suzy became the wife of bandsman Robert Draho. It should be noted that some of the dancers were also musicians, and often performed in that capacity with the Band, although in competitions they could only compete as either a dancer or a musician, never both.

The Band kept busy throughout the 1980s, beginning the new decade with a March parade commemorating the Commonwealth Wartime Aircrew Reunion being held in

MCPL DON BLAINE PIPES AT THE COMMISSIONING OF *HMCS WINNIPEG* IN ESQUIMALT, BC, ON 23 JUNE, 1995

Winnipeg. A year later, they played a Winnipeg Concert Hall Performance to raise funds for blind children; an event sponsored by the Honourable F.L. Jobin, then-Lieutenant-Governor of Manitoba, in association with the Military

Police.[8] The CFB Portage la Prairie Pipe Band, the 2nd Battalion PPCLI Drum Corps from Winnipeg, and the Air Command Band, all took part under Capt Keith Swanwick, the Director of Music for the Air Command Band. July found the Pipes and Drums in Calgary performing as part of a Manitoban Massed Military Pipe Band in the 1981 Calgary Stampede, along with the Air Force Pipe Band from Portage la Prairie and the Cameron Highlanders. The Band was awarded first place in the parade. Later that day, they presented a display on the Stampede's Main stage.

THE BAND POSES ON A SCOTTISH HILLSIDE LOOK-A-LIKE

1982 was another important year, as brothers Pipe Major Neil and WO John Barbour accompanied the Air Command Band to play in ceremonies at Malmstrom Air Force Base in Great Falls, Montana. They were part of the commemoration of the 25th Anniversary of the North American Aerospace Command and Canadian Forces Association with the Malmstrom Base in June of 1982. On 31 July, the Band performed at the ceremonial presentation of colours to Air Command. A few years later, in 1984, they performed at the Royal Manitoba Festival in honour of the Royal Visit, receiving a special thank-you from Manitoba Premier Howard Pawley:

> I know Her Majesty enjoyed her (sic) whole stay in Manitoba but I am sure the highlight of Her visit was the Royal Manitoba Festival. Her obvious delight in watching the performance was shared by all in the Royal Box and most definitely by my wife and myself.
>
> I would like to take this opportunity to thank you for your participation in an evening that was truly a tribute to Manitoba.

The Pipes and Drums continued with yet another excellent concert in 1988 at the Nova Scotia Tattoo in Halifax.

It was not only the pipers and drummers who consistently won awards and recognition during their many competitions over the years, so too did the Highland dancers. One dancer who stood out in the earlier years was Jillian Currie. Her credentials were world-class: she was the Canadian Highland Dance Champion of Canada in 1988-1989

JILLIAN CURRIE, THE CANADIAN HIGHLAND DANCE CHAMPION OF CANADA IN 1988-1989

AN IMAGE THAT APPEARED IN "SENTINEL" MAGAZINE SHOWS THE BAND READY TO "STRIKE UP" AT EPCOT CENTRE, WALK DISNEY, FLORIDA, 1989

the Epcot Centre in a 'walkabout' (a non-static performance where they played while marching through the park with no specific destination). In return, once the performance was complete, they were given the run of the park.

and silver finalist at the World International Competition in Dunoon, Scotland.

402 helped augment the Portage la Prairie Pipe Band at the Strawberry Festival in Plant City, Florida, in 1989. Plant City and Portage were the Strawberry Capitals of Florida and Canada respectively, and thus became twinned cities. While in Florida, the Band performed at Walt Disney World and

They played at the Long Peak Highland Festival in Estes Park, Colorado on 8-9 September 1990. As a result of their concert, Major Bob Tracey, a Canadian working at NORAD in Colorado Springs, commissioned a special banner for the Band in recognition of the wonderful show they had presented. The banner, with the Air Command Crest on one side and the 402 Squadron Crest on the other, was stitched with gold metallic thread.

As the guest band at the Rio Grande Valley Celtic Festival in Albuquerque, New Mexico, 15-18 May 1992, the Band led the parade in the opening and closing ceremonies. Several members won individual competitions: Pte Sean Osztian and Cpl James Symonds (former CO, LCol Symonds' son) each won silver in the drumming competitions; Cpl Douglas Knight won the gold medal in Piping; and Pte K.L. MacDonald won the gold medal in all five of her dance competitions. MacDonald, who went on to win the Jack Reeve Award for the Squadron's Airwoman of the year, was also named Dancer of the Day, and presented with two additional trophies for her feats. The performance by the Band and its dancers made such an impression on both sides of the border that CFB Winnipeg's newspaper, the *Voxair*, printed the event organizers' letter of appreciation to the Squadron CO, LCol R.W. Patrick, in its entirety.[9]

Bagpipes were – and still are – considered instruments of war (German troops were unnerved by the unnatural sound in WWI).[10] With this in mind, Pipe Major Neil Barbour's presence was requested at the Readiness Challenge IV in April, 1993, at Elgin Air Force Base. His task was to lead the Canadian contingent into the competition and act as a moral booster for the team. This was the first time a military force from outside the United States competed in the grueling week-long challenge. One competition, called 'The Fog of War', involved a nighttime obstacle race through firing bullets and mortars. As the Canadian contingent came over the hill for the last stretch, Barbour stepped out from the shadows and piped them to a winning finish.

Eventually, the Governor of Texas declared the Band and its members honourary citizens of the State of Texas. The seeds of friendship sown during these performances eventually led to the twinning of Dallas Naval Air Station and 402 as sister squadrons. In June 1994, Pipe Major Barbour and Squadron officers traveled to Dallas Naval Air Station to take part in numerous 50th anniversary ceremonies in remembrance of D-Day.

The 402 Band has always performed to the highest standards, leaving lasting impressions on all they meet, particularly in Texas. Capt McFee from the 19th Regiment in Texas of the Texas State Guard, acted as liaison bringing the 402 Pipe and Drum Band to the State and helping to escort them during their trips to the Texas Scottish Festival and Highland Games in Dallas, where they performed several times. McFee wrote in his 1994 letter of appreciation to Prime Minister Jean Chrétien:

Dear Honored Sir:

It is with the most profound sense of respect and humility that I take the occasion to bring to your attention the wealth of honor and respect the nations of Canada has in the women and men of the Royal Canadian Air Force Military Pipe Band, SQD 402, WING 17, City of Winnipeg. The 1/19th REG was given the distinction of escorting this unit and then watched in amazement as they, in 3 days, managed to conquer the hearts of every Texan they met.

Sir, I have served my nation as a Reconnaissance Marine and unfortunately, endured the stark rigors of war. I was blessed with extensive training and with crack personnel to command yet, in all my experience, I have never had the privilege to share in the overwhelming Canadian professionalism, to soar the heights of Canadian honor and to proudly observe the profound respect they commanded from every Texan they met.

I cannot begin to convey what this unit has meant to me and to the members of my command. The women and men of the 402nd deserve the highest recognition as ambassadors of the nation they love. They came as the best Canada has to offer and took our hearts with them. Prime Minister, I cannot endure the loss, for well-worn memories will not satisfy the pain of separation. They must come back to Texas.[11]

He wrote again that year to the editors of the *Winnipeg Free Press*, and then again three years later, expressing his heartfelt admiration for the Pipe Band of the 'Bear Squadron':

O Canada, we wept as they left – we always weep, for they have captured our respect, our admiration and our hearts.

It is extremely difficult to stand on the tarmac and watch a Dash-8 lift into the unforgiving, hot, humid Texas sky knowing that the best Canada has offered will not be back for another year to grace the Texas Scottish Festival. It is hard, so very hard.

You do not know just how much 402 Squadron means to us. Four years ago, the 402 pipe band played its way into our hearts, and each time it returns, the members reaffirm everything we knew from the beginning: they represent the best Canada has to offer.

But do not ask me to confirm this; ask the thousands of Texans who gave a standing ovation at their performances; ask the U.S. navy, which consistently provides billeting and transportation as a "sister unit"' ask the men of the Texas State Guard who volunteer to support their every need.

Canada, we would march into hell and shout at the devil for the women and men you sent as ambassador to our state. Cherish what you have; give them the honour they so richly deserve.[12]

It may seem to some that being on the road is not really an onerous task, and one that could be done well without too much effort. However, at these performances, eyes are on the Band members, and fair or not, their actions are under constant scrutiny. The Pipes and Drums have always been aware that they are ambassadors for the Squadron, the Air Force, the City of Winnipeg and of course, Canada. Being no strangers to the road, they remain conscious of life under a microscope, and conduct themselves as the honoured guests they are. In fact, this ambassadorial skill has become a very important part of what they do. Perhaps Band pride has never been tested to the extent previously noted in Cpl MacLeod's commendation for his performance while injured. After joining the Band on a trip to perform at the University of Texas, Maverick Stadium in 1995, Cpl Ken Allen wrote: "A typical day on the road begins at seven am and usually ends sometime early the next morning. After the official performances they bring their bagpipes and drums everywhere they go and will play for anyone, anytime."[13] Cpl Allen, a military photographer, accompanied them on a CC-142 Dash-8 to Naval Air Station (NAS) Dallas, Texas and wrote of the great pride he felt while watching their show, both at events and as ambassadors with the Texan public. Their friendly attitude, combined with the willingness to answer any question and pose for photo after photo, impressed him; but it was their outstanding work ethic and pride in uniform that left a bigger impression. While in Dallas, the Band was also invited to visit the new Texas Ranger Stadium in Arlington.

THE BAND PERFORMING IN DALLAS, TEXAS ON A WARM DAY IN 1995

Shortly thereafter, the Band performed at an event of particular significance, not just for the Band but also the City of Winnipeg. On 23 June 1995, the 402 Squadron Pipes and Drum Band were invited to Esquimalt to perform at the commissioning of *HMCS Winnipeg*. Then-Mayor, Her Worship, Susan Thompson, as well as LCol McIntyre, CO of 402, were also present for the happy occasion. During the ceremony, a Sea King and a Dash-8 both carried out a fly-past over the ship. After the ceremonies were over, the Band, along with guests and media went on a short cruise in the newly commissioned Halifax-class vessel.

A "CITY OF WINNIPEG" SQUADRON DASH-8 OVER-FLIES *HMCS WINNIPEG*. AS THE SQUADRON AND SHIP TEAMED UP IN 1995, THE ASSEMBLED WERE TREATED TO A STIRRING CONCERT FROM THE BAND

MWO Neil Barbour was honoured with an invitation to perform at a memorial service in Alaska on 27 September 1995. The service commemorated twenty-four aircrew from E3 Flight YUKLA 27 of 962nd ACCS Flights. Their plane had crashed on a routine mission with no survivors. The letter from Senior Canadian Officer at Elmendorf Air Force Base, Alaska, to the Commander of Air Command Headquarters commended MWO Barbour for his touching performance:

> As the first note of *Amazing Grace* lofted in the air at Hangar 1, the overwhelming sense of pride and the bond among the Detachment members and their families was evident. Master Warrant Officer Neil Barbour performed with dignity and provided inspiration and the Canadian influence on the solemn occasion marking the passing of Sergeant David Pitcher and Master Corporal J. Legault and their fellow American aircrew members.[14]

MWO NEIIL BARBOUR ACCEPTING CONGRATULATIONS

PIPE AND DRUM BAND

THE BAND IN BONNETS

THE BAND IN FLIES AND TIES

As the 1990s came to a close, the Band did not slacken its pace. Its activities continued into the millennium, despite spending reductions, loss of Squadron transport capabilities with the sale of the two CC-142 transport Dash-8s, and a continuing battle for a permanent space to practice.

The Band's performance may have hit a high mark during the Squadron's the 70th Anniversary in October 2002, when they delivered a rendition of *Amazing Grace* so stirring that many Squadron vets, including the late D.B. Riddell, a Second World War Spitfire pilot who attended in a wheelchair, openly wept. Such is the power of their piping.

One of the highlights for the Band in the new century must surely be the winning of a coveted place with the Ceremonial Guard on Parliament Hill in 2003. Pte K.L. MacDonald, a university student and a City of Winnipeg Lifeguard, was also a former Highland Dancer with the Band. She had hung up her shoes due to an injury and turned to her other love, the bagpipes. Kirsten, along with 200 other applicants, competed for the eighty annual summer positions in the composite

Army Reserve unit. Her duties that summer included playing at the daily Changing of the Guard ceremony, mounting sentries, and performing at official Rideau Hall functions, foreign embassies and special events for dignitaries.

Recently, Band members have won back to back honours winning the prestigious Jack Reeve Memorial Air Reservist of the Year Award two years in a row. Bill Crosby won in 2005, and current Band Leader, MCpl Greg McTavish, was airman of the year for 2006. Both men were recognized for their tireless efforts in promoting the Band and the Squadron.

The Band continues to perform at many civic and military functions throughout the city and the province, especially the air shows. They were recently part of the opening ceremonies for the *Run for the Cure* breast cancer fundraiser, and annually lead the Squadron at the Remembrance Day ceremony at Bruce Park and the subsequent march down Portage Avenue to the St. James Legion.

Despite their ceremonial function, the Pipes and Drums (like all military bands and especially pipe bands) serve as a public face that instills pride and admiration for the Canadian Forces. Over the years, the Band has brought millions of smiles to people all over the world, moved countless others to tears and captured the hearts of all who attend their concerts. As Winnipeg Bear Ambassadors, they are a priceless part of 402 Squadron.

THE INDOMITABLE CLIFFORD COOKE IN FULL DRUM MAJOR REGALIA

NOTES

1. Neil Barbour, "History of Royal Canadian Air Force: #402 Sqdrn. (Aux) Pipe Band, Winnipeg, Man.", pg 3, 1981. 402 Sqn Archives, Band Files.
2. Barbour, Ibid, 2.
3. Interview with John Reay Jr., October 2006.
4. Barbour interview, December 2006.
5. 402 Sqn Archives, clipping record book.
6. 5050-1 (Bcomd) 30 May 1978, 402 Squadron Archives.
7. *HMCS Chippawa* is named after *HMS Chippewa* and sailed under Commander Robert H. Barclay, whose fleet operated on Lake Erie during 1812-1813 – The Insignia and Lineages of the Canadian Forces Volume 2, Part 1, Extant Commissioned Ships.
8. *Voxair*, 12 March 1980.
9. *Voxair*, 22 July 1992.
10. Neil Barbour interview, 16 Jan 2007.
11. DND File No. 9404199 4 Aug 1994, 402 Sqn Archives.
12. McAffee, Michael L., Grand Prairie, Texas, *Winnipeg Free Press*, Friday, 20 June 1997, page A13, Letters to the Editor.
13. *Voxair*, 21 June 1995.
14. 1045-1 (ANR/CD) 29 Sept 1995.

Appendix A

112 Squadron Officers – Known Commands and Honours

This is not a complete list of officers. Initial rank is that granted upon enrolment in 112 Squadron. Appointments are not necessarily the last held.

NAME	HIGHEST RANK AND DECORATIONS		HIGHEST APPOINTMENT
S/L J.A. Sully	A/V/M	CB, AFC	Air Member for Personnel
F/L H.P. Crabb	W/C		Director of Manning
F/L J.C. Huggard	S/L		CO 1 (Conversion) Sqn
F/L R.H. Little	W/C		
P/O R.J. Clement	W/C	AFC	CO 402 (Fighter) Sqn
P/O E.R. Gardner	S/L		CO 121 (Composite) Sqn
P/O E.H. G. Moncrieff	G/C	AFC	OC 39 (Recce) Wing
P/O V.H. Patriarche	A/C	OBE, AFC	Comd Northwest Staging Rte
P/O W.W.S. Ross	S/L		CO 123 (Army Co-op) Sqn
P/O G.H. Sellers	G/C	AFC	OC 39 (Recce) Wing

Two 112 Squadron officers are known to have died on active service. Rank given is the rank held at death.

F/L A.H.F. Alloway – Killed in action 22 January 1941

F/L A.B. Jobin – Killed in a flying accident 11 November 1941 while flying Lysander 423, 1 (Coastal Artillery Co-operation) Squadron, St. John

Appendix B
Squadron Commanding Officers

S/L J.A. Sully, AFC	01 Mar 1933 - 30 Sep 1937	
S/L H.E. Crabb	01 Oct 1937 - 30 Sep 1939	
S/L R.H. Little	01 Oct 1939 - 14 Apr 1940	
S/L W.F. Hanna	15 Apr 1940 - 06 Jan 1941	
S/L G.R. McGregor, DFC	07 Jan 1941 - 13 Apr 1941	
S/L V.B. Corbett	14 Apr 1941 - 13 Dec 1941	
S/L R.E.E. Morrow, DFC	14 Dec 1941 - 16 Aug 1942	
S/L N.H. Bretz, DFC	17 Aug 1942 - 26 Sep 1942	
S/L D.G. Malloy, DFC	27 Sep 1942 - 14 May 1943	
S/L L.V. Chadburn, DFC	15 May 1943 - 12 Jun 1943	KIA
S/L P.L.I. Archer, DFC	13 Jun 1943 - 17 Jun 1943	KIA
S/L G.W. Northcott, DSO, DFC and BAR	18 Jun 1943 - 28 Jul 1944	
S/L W.G. Dodd, DFC	29 Jul 1944 - 28 Oct 1944	
S/L J.B. Lawrance	29 Oct 1944 - 21 Feb 1945	
S/L L.A. Moore, DFC	22 Feb 1945 - 25 Mar 1945	KIA
S/L D.C. Laubman, DFC and BAR	06 Apr 1945 - 14 Apr 1945	MIA
S/L D.C. Gordon, DFC and BAR	15 Apr 1945 - 10 Jul 1945	
W/C R.J. Clement, DFC	02 Aug 1946 - 21 Dec 1948	
W/C L.M. Cameron, DFC	22 Dec 1948 - 27 Feb 1950	
W/C W.B. Breckon	28 Feb 1950 - 07 Nov 1951	
W/C D.W. Rathwell, DFC	08 Nov 1951 - 07 Oct 1953	
W/C J.M. Reid, CD	08 Oct 1953 - 14 May 1956	
W/C D.M. Gray, CD	15 May 1956 - 17 Jul 1960	
W/C J.T. Patterson, CD	18 Jul 1960 - 15 Oct 1962	
W/C D.R. Scott, CD	16 Oct 1962 - 15 Oct 1965	
W/C J.A. Brown, CD	16 Oct 1965 - Jun 1971	
LCol E.J. Harris, CD	Jul 1971 - Sep 1971	
LCol R.D. Wilson, CD	Sep 1971 - Dec 1973	
LCol E.T. Wagner, CD	Dec 1973 - Jun 1975	
LCol J.C. Haip, CD	Jun 1975 - Dec 1977	
LCol R.W. Slaughter, CD	Dec 1977 - 31 Oct 1981	
LCol M.S. Joyce, CD	01 Nov 1981 - Mar 1983	
LCol L.E. Olsen, CD	Mar 1983 - May 1986	
LCol J.M. Symonds, CD	May 1986 - Jun 1988	
LCol R.W. Patrick, CD	Jun 1988 - Aug 1992	
LCol C.V. MacIntyre, CD	Aug 1992 - Sep 1995	
LCol P.G. Rawlings, CD	Sep 1995 - Jan 1997	
LCol D. Lamb, CD	Jan 1997 - May 2000	
LCol B. Doyle, CD	May 2000 - Jul 2003	
LCol S.L. Schock, CD	Jul 2003 - Jul 2006	
LCol R.T. Witherden, M.B., CD	Jul 2006 -	

Index

1st **Canadian Division** 24

2nd **Armoured Car Regiment** 18

3rd **Regiment, Royal Canadian Horse Artillery** 160

8th **Air Force (U.S.A.)** 58, 59, 62, 63

29th **Air Division (U.S.A.)** 127

3 **Air Reserve Region Headquarters** 173

3 **Air Reserve Western Headquarters (3 ARWHQ)** 173

3 **Rescue Support Unit (3 RSU)** 173, 174

4 **Wing Cold Lake** 194

6 **Division Royal Canadian Army Service Corps** (RCASC) 117

10 **Tactical Air Group (10TAG)** 190

11 **Technical Servicing Unit** (11 TSU) 107

11C **mobile radar convoy** 104

17 **Wing Winnipeg** (see also No.17 Wing) 126, 146, 148, 149, 151, 154, 156, 158, 192-194, 201

18 **Wing Edmonton** 119

400 **Squadron** 190, **198** *notes*, 202

401 **RCAF Squadron** 59

402 **Scramble Plan** 148

402 **Technical Training Unit** (402 TTU) 115, 185

403 **RCAF (Calgary) Squadron** 60, 87, 109, 119, 125, 127, 128

406 **(Saskatoon) Squadron** 108, 109, 119, 120, 125

411 **Squadron** 87

412 **RCAF Squadron** 59, 90

418 **(Edmonton) Squadron** 108, 109, 120, 124, 125, 127, 128, 155, **161** *notes*, 172

416 **RCAF Squadron** 63, 65

417 **Squadron** 87, 104, 109

421 **Squadron** 63

425 **(Bagotville) Squadron** 178

429 **Squadron** 176, 185

441 **RCAF Squadron** 75

442 **(Vancouver) Squadron** 87, 109, 119, 122

2402 **Aircraft Control and Warning Unit (2402 AC&WU)** 115, 116, 120, 121, 125, 127, 129, 130, 133, 140, 141, 149, 179

3052 **Technical Training Unit (3052 TTU)** 115, 116, 126, 141, 149, 153

4003 **Intelligence Unit** 140

4003 **Maintenance Unit** (4003 MU) 115, 151, 153

4003 **Medical Unit** 144, 145

5001 **Intelligence Unit** 115, 116, 143

No.1 **Group** 38

No.2 **Air Command** 98, 102

No.2 **Air Liaison Section (Army)** 36

No.2 **Canadian Squadron** 32, 33, 56

No.2 **Group** 42, 53, 55

INDEX

No.3 All-Weather (Fighter) Operational Training Unit 136

No.3 Group 69

No.8 Group 69

No.10 Squadron (later No.110 Squadron) 30

No.11 Squadron (later No.111 Squadron) 14, **28** *notes*

No.11 Group 41, 52, 53, 61, 108, 119

No.12 Group 33, 34, 61

No.17 Wing (see also 17 Wing Winnipeg)

    and (Auxiliary) 120

    and (Reserve) 120

No.17 Armament Practice Camp (No.17 A.P.C.) 82

No.65 Squadron 39

No.83 Group 83, 90

No.83 Ground Support Unit 78

No.85 Group 68

No.102 (Army Cooperation) Wing 22

No.125 (RAF) Wing 75, 78, 79

No.126 Wing 81, 90

No.127 (RCAF) Wing 78, 79, 90

No.142 (RAF) Wing 68

No.222 Squadron 42

No.234 Squadron 39, 44, 49

No.409 Repair and Salvage Unit 80

No.410 Squadron 101

No.501 Squadron 44, 49

No.6402 Squadron 75

No.6441 Squadron 75

Ainley, R.B., Major (Maj) 175

Air Command 17, 23, 98, 103, 106, 110, 111, 114-116, 119, **162** *notes*, 178, 181, 185, 192, 208, 209

    and Headquarters 179, 213

Air Defence Command 128, 130, 132-134, 138, 140, 145

Air Defence Group 108

Air Defence of Great Britain (A.D.G.B.) 68, **92** *notes*

Air Force Auxiliary 1, 5, 115

Air Force Cross 5

Air Force Day 121, 128, 133, 136, 139, 144

Air Force Headquarters 11

Air Ministry Experimental System (AMES) 104, 116

Air Transport Command (ATC) 108, 109, 146, 149, 152, 156

    and Air Standards Unit 159

Allen, Ken, Corporal (Cpl) 212

Anderson, Tommy, Squadron Leader (S/L) 106

Archer, P.L.I., Squadron Leader (S/L) 63, 64, 218

Ardennes Offensive 81

Arnhem (Holland) 76, 86, 90, 95

Ashley, L.A., Lieutenant-General (LGen) 181

Atholl Highlanders March 200

Austin, W.G., Flight Sergeant (FSgt) 73

Auxiliary Active Air Force 22

Aviation League of Canada 4

Ayr (Scotland) 41

B.116 Wunstorf 86, 87

B.152 Fassberg 90

B.70 Antwerp 75

B.64 Diest 79

B.78 Eindhoven 79

B.80 Volkel 82

B.82 Grave 75, 76, 78

B.108 Rhine 86

Bailey, D., Lieutenant (Lt) 171

Barber, V.E., Warrant Officer First Class (WO1), Flying Officer (F/O) 88, 90, 112

Barbour, John, Warrant Officer 202, 208

Barbour, Neil, Warrant Officer (WO), Master Warrant Officer (MWO), Pipe Major 205-207, 210, 213, **216** *notes*

Baril, M., General 191

Barnard, A.E., Flying Officer (F/O), Flight Lieutenant (F/L) 78, 80, 81

Base Aircraft Engineering Organization (BAMEO) 187

Bastable, V.J., Flying Officer (F/O) (see also Vernon Bastable Memorial Award) 106-108, 170, 171, 194

Battle of Britain 8, 31, 33, 34, 42, 55, **92** *notes*, 125, 128, 135

Battle of Britain Sunday 128, 135

Battle of France 29

Battle of the Bulge (see also Ardennes Offensive) 80

Bavis, C.H., Pilot Officer (P/O) 69, 74

Bayeux War Cemetery, France 71

Bayly, J.C.U., Pilot Officer (P/O) 36

Bell-Irving, A.D., Major (Maj) 4

Bell, J.A., Corporal (Cpl) 192

Berlin Airlift 111

Berlin War Cemetery, Charlottenburg, Germany 87

Bird, K.W., Flight Sergeant (FSgt) 36, 63

Birt, Rob (Squadron groundcrew) 62

Bissky, P., Flight Lieutenant (F/L) 99

Blaine, Don, Pipe Major 207

Bland, E.A., Flight Lieutenant (F/L) 57

Bourgeon, Ray 31

Braceland, Hugh, Squadron Chief Engineering Officer 129

Breadner, Air Commodore (A/C) 22, 23

Breckon, W.B., Wing Commander (W/C) 112, 120, 165, 218

Bretz, N.H., Pilot Officer (P/O), Flight Lieutenant (F/L), Squadron Leader (S/L) 36, 44, 49, 55, 58, 62

British Commonwealth Air Training Plan (BCATP) 1, 55, 56, 79, 98, 103, 178

British Expeditionary Force 29

Bromley Observers' Post 73

Brookwood Military Cemetary, Woking, Surrey 49

Brown, Sergeant (Sgt) 16

Brown, J.A., Wing Commander (W/C) 158, 218

Brown, O.R., Flight Sergeant (FSgt) 36, 60, 63

Burrows, E.R., Flight Lieutenant (F/L) 85, 88, 89

C. De la Hague 45

Calais 42, 47, 58

Campbell, R., Flight Lieutenant (F/L), Band President/Band Officer 200

Cameron, G.D., Sergeant (Sgt) 36

Cameron, L.M., Sergeant (Sgt), Flying Officer (F/O), Wing Commander (W/C) 36, 60, 61, 106, 107, 112, 218

INDEX

Camp Borden 9, 15, 124

Camp Gagetown 154

Camp Shilo 7, 10-17, 120, 150, 151, 157, 159, 160

Canadian Fighter Pilots' Association 179

Canadian Flying Clubs Association 6

Canadian Forces Air Navigation School (CFANS) 180, 184, 192, 196

Canadian Forces Base (CFB) Edmonton 173

Canadian Forces Base (CFB) Petawawa 174

Canadian Forces Base (CFB) Winnipeg 173

Canadian Forces Base (CFB) Toronto 159, 160

Canadian Forces Europe (CFE) 205, 206

Canadian Joint Air Training Centre (CJATC) (see also Rivers) 108, 109, 115, 125, 151

Cannon, A.E., Pilot Officer (P/O) 26

Cantrill, C.T., Flying Officer (F/O) 36

Carpenter, Graham 34

Central Air Command 102

Central Flying School (CFS) 114, 140, 144, 186, 188, 195

Chadburn, L.V. "The Angel", Pilot Officer (P/O), Squadron Leader (S/L) 62, 66, 218

Chalkley, Al (Squadron airman) 124

Chambers, Jennifer, Corporal (Cpl) 189

Chandler, R.E., Pilot Officer (P/O) 27

Channel Stop 43, 44, 92 *notes*

Chapman, S.J., Lieutenant-Colonel (LCol) 180

Charbonneau, P., Corporal (Cpl) 192

Chase, H., Captain (Capt) 186

Chief of Air Staff 25, 196

Churchill, Winston 31, 58

Classon, Nicholas W. 205

Clayton, Diane, Master Corporal (MCpl) 189

Clayton, Ron, Major (Maj) 174

Clement, R.J., Wing Commander (W/C) 98, 104, 217, 218

Cleverley, Fred 178

Cold War 2, 102, 110, 120, 131, 137, 143, 169, 178, 184

Collette, D., Officer Cadet (OCdt) 181

Collins, K.M., Flying Officer (F/O) 71

Commanding Officer's Commendation 191

Compton, H.N., Corporal (Cpl) 187

Cooke, Cliff, Drum Major 199, 205, 206

Corbett, V.B., Flight Lieutenant (F/L), Squadron Leader (S/L) 36, 38, 39, 41, 42, 44, 45, 218

Cosme, Joe, Sergeant (Sgt) 186

Cowan, H., Flight Lieutenant (F/L) 85, 87

Crabb, H.P., Squadron Leader (S/L) 3, 6, 22, **28** *notes*, 217, 218

Craig, Andrew, Aircraftman Second Class (AC2) 7

Crease, H.S., Pilot Officer (P/O), Flying Officer (F/O) 36, 38, 42

Croil, G.M., Air Vice Marshal (A/V/M), Air Marshal (A/M) 14, 22, 25

Crosby, Bill (Bandsman) 215

Cummings, Keith, Lieutenant (Lt) 205

Cunniff, B. "Brian", Major (Maj) 175

Currie, Jillian (Dancer) 208, 209

D-Day 62, 68, 69, 80, 193, 210

Damery, Stephen, Air Cadet 183

Das, Tulse, Corporal (Cpl) 173

Dash-8 Master Implementation Plan (MIP) 181

Davidson, Bill (Squadron airman) 129

Davidson, Mr. John 178

Deelen Airfield 86

Defence Planning Guidance 2000 192

Demare, Flight Lieutenant (F/L) 160

Dempster, Jack, Flight Lieutenant (F/L) 106, **161** *notes*

Den Helder (Netherlands) 67

De Niverville, J.A., Flight Lieutenant (F/L) 73

Department of National Defence 4, 5, 147, **161** *notes*

DesAutels, B.P., Sergeant (Sgt) 175

Desmedt, Peter, Brigadier-General (BGen) 190

Dew, R.J., Flying Officer (F/O) 119

Dieppe 56, 57, 90, **92** *notes*, 95

Digby (See also RCAF Station Digby) 33-38, 40, 41, 61, 64, 66, 67, 81

Distinguished Flying Cross (DFC) 8, 34, 45, 49, 53, 55, 59, 63, 64, 65, 67, 68, 71, 82, 84, 86, 87, 90, 95, 98, 107, 112, 120, 129, 165, 179, 201, 218

Dodd, W.G., Flight Lieutenant (F/L), Squadron Leader (S/L) 68, 71, 218

Dodmans Point 48

Domini, Benny (Squadron groundcrew) 62

Draho, Robert (Bandsman) 207

Doyle, B.M., Lieutenant-Colonel (LCol) 197, 218

Drummond, D.R., Flight Lieutenant (F/L) 89

Dube, Richard, Sergeant (Sgt) 189

Duguid, Jim, Leading Aircraftman (LAC) 27, 34

Dupuis, J.G., Sergeant (Sgt) 175

Dutton, H.G., Flying Officer (F/O) 88

E-boat 61, 63

Eady, I.J., Sergeant (Sgt) 49

Eastern Air Command 101

Edwards, D.M., Flying Officer (F/O), Flight Lieutenant (F/L) 3, 8

Eindhoven 75, 79, 80, 81

Elliot, L.G., Flight Sergeant (FSgt) 51

Emberg, Flight Sergeant (FSgt) 48

Engleberg, Madeline (Dancer) 207

Engleberg, Suzy (Dancer) 207

English Channel – See Channel 30, 90, **92** *notes*, **93** *notes*, 95

Enright, Warrant Officer (WO) 179

Evans, E.H., Flight Lieutenant (F/L) 17

Exercise Abegweit I 174

Exercise Assiniboine 115, 120, 121, 126

Exercise Buffalo 106

Exercise Buffalo IV 128

Exercise Checkpoint 134

Exercise Cottontail 130

Exercise Cross Border 135

Exercise Eagle 66, 110, 119

Exercise Goldeye 141

Exercise Lynx-Cat 129

Exercise Nugget 121, **162** *notes*

Exercise Panti-Hose 174

Exercise Rendezvous 43, 62, 179

   Rendezvous '85 179

   Rendezvous '87 179

Exercise Strike 120, 125

Exercise Tailwind 128, **162** *notes*

INDEX

Exercise Tempered and Polished Bayonets 160

Exercise Windmill I 171

Exercise Windmill II 171

Fallis, Bob (Squadron airman) 31

Fee, J.C., Wing Commander (W/C) 59

Ferwerda, J.N., Corporal (Cpl) 175

Fidelak, Leading Aircraftman (LAC) 130

Findlay, H.J., Flying Officer (F/O) 36

Finucane, B., Wing Commander (W/C) 55

First Canadian Army 114

Firth, W., Sergeant (Sgt) 148

Ford, L.S. Pilot Officer (P/O), Flying Officer (F/O), Squadron Leader (S/L) 44, 45, 48, 60

Foster, F.B., Pilot Officer (P/O), Flying Officer (F/O) 36, 41, 44

Franco-British War Cemetery, St. Valery-en-Caux 57

Frederick, John, Flight Sergeant (FSgt), Assistant Band Master 200, 202

Fry, M., Corporal (Cpl) 177

Gates, "Doc", Flight Lieutenant (F/L) 68

Gates, Moe, Colonel (Col) 171

Gayfer, Bud 31

Gilland, Frank, Flying Officer (F/O) 128

Gillespie, R.R., Pilot Officer (P/O) 36

Gimbel, E.L., Pilot Officer (P/O) 60

Globe and Mail 173

Gobeil, F.M., Flying Officer (F/O) 8

Gordon, D.C., Squadron Leader (S/L) 34, 87, 88, 89, 218

Gordon, J.L., Group Captain (G/C) 5, **28** *notes*

Graham, L.R., Corporal (Cpl) 41

Grave (Holland) (see B.82 Grave) 75, 76, 79

Gray, D.M., Squadron Leader (S/L), Wing Commander (W/C) 106, 134, 135, 139, 144, 148, 151, 202, 218

Griffiths, Owen (Squadron airman) 27

Guthrie, K.M., Flight Lieutenant (F/L), Air Vice Marshal (A/V/M) 6, 28, 106

Hagenow Aerodrome 88

Hall, Jackie 173

Hamilton, M.G., Corporal (Cpl) 191

Handley, K.B. "Butch", Sergeant (Sgt) 36, 37, 41, 44

Hanna, W.F., Squadron Leader (S/L) 32, 218

Hardy, Sergeant (Sgt), Flying Officer (F/O) 17, 103

Harris, E.J. "Ernie", Wing Commander (W/C), Lieutenant-Colonel (LCol), Colonel (Col) 147, 148, 152, **162** *notes*, **163** *notes*, 171, 173, 218

Harvey, W.S., Flight Lieutenant (F/L) 83

Hawkinge – See RAF Station Hawkinge 75

Heakes, E.V., Group Captain (G/C) 31

Hebert, R., Captain (Capt) 183

Heesch 81, 82, 88

Henderson, Al, Flying Officer (F/O) 145, 151, 157, **163** *notes*

Higgins, Mush (Squadron airman) 31

HMCS Bonaventure 188

HMCS Chippawa 177, 206, **216** *notes*

HMCS Shearwater 120

INDEX

HMCS Winnipeg 190, 207, 213

HMS Chippewa 216 notes

HMS Victory 177

Ho Ming 5 197

Hodson, H.L.B., Wing Commander (W/C) 60

Honeycombe, R.B., Sergeant (Sgt) 58

Hoppensack, M.A., Private (Pte) 186

Houston, D.C. "Dave", Lieutenant (Lt), Major (Maj) 174, 175

Hubbard, K.S., Corporal (Cpl) 192

Hudson, Jack, Squadron Leader (S/L) 106, **161** notes

Huggard, Flight Lieutenant (F/L) 6, 7, 217

Hyde, G.G., Flying Officer (F/O) 36, 39

I./KG 51 (German Squadron) 84

Innes, B.E., Flight Lieutenant (F/L) 50, 61, 82, 83, 85, 89

International Civil Aviation Organization (ICAO) 103, 144

Irvine, K.H., Flying Cadet 142

Jack Reeve Memorial Award 173, 188, 189

James, R.G., Honourary Colonel 158

James, W.R. 106

Jenkin, D.W., Sergeant (Sgt) 36, 42

Jobin, F.L., Lieutenant Governor of Manitoba 207, 217

Jobin, Marcel 25, 31

Johannason, Connie, Chief Instructor 8

Johnson, Art 6, 24, **28** notes

Johnson, G.O., Group Captain (G/C) 17

Johnson, R.G., Squadron Leader (S/L) 107

Joint Air School (JAS) (see also Rivers) 104, 105, 109

Jones, G., Sergeant (Sgt) 177

Joyce, M.S. "Malcolm", Lieutenant-Colonel (LCol) 174, 175, 177, 218

Keene, N.A., Flight Sergeant (FSgt), Flying Officer (F/O) 36, 48, 61

Kelly, F.W., Pilot Officer (P/O), Flying Officer (F/O) 36, 43

Keltie, I.G., Pilot Officer (P/O) 36, 53, 57

Kenley Wing (see also RAF Station Kenley) 57

Kieth, G.N., Pilot Officer (P/O) 36, 60

King, Mackenzie, Prime Minister of Canada 97

Klaponski, Frank, Corporal (Cpl), Flight Sergeant (FSgt) 6, 7, 24, 34, 101

Knight, Douglas, Corporal (Cpl) 210

Lacoursiere, G.J.R., Corporal (Cpl) 187

Lamb, D.G. "David", Lieutenant-Colonel (LCol) 190, 218

Langruth air-to-ground firing range 125

Land Forces Command 189

Laubman, D.C., Squadron Leader (S/L) 86, 90, 218

Lawrence, George, Warrant Officer (WO), Pipe Major, Band Officer 205, 206

Lawrence, J.B. "Bud", Flight Lieutenant (F/L), Squadron Leader (S/L) 76, 79

Lawson, R.W. "Bob", Flying Officer (F/O) 80, 85

Leask, H., Private (Pte) 177

Lee, T.B., Flying Officer (F/O) 88

Lee, W.R., Squadron Leader (S/L) 116

Linden, Eric W., Major-General (ret'd) 2, 192

INDEX

Little, R.H., Flight Lieutenant (F/L), Squadron Leader (S/L) 6, 24, 28, 217, 218

Little, T.B., Flying Officer (F/O), Flight Lieutenant (F/L) 26, 42

Lloyd, G., Corporal (Cpl) 177

Lockhart, P., Corporal (Cpl) 177

Lord Rothchild 31

Lord Tweedsmuir, Governor-General 26

Low, Sergeant (Sgt) 130

"Maasburg" 67

MacBrien Trophy Shoot 134

MacDonald, K.L. Private (Pte) 189, 210, 214

MacIntyre, C.V. "Chuck", Captain (Capt), Lieutenant-Colonel (LCol) 177, 188, 218

MacKay, G.P., "Jerry", Pilot Officer (P/O) 57

MacLauchlin, Smokie (Squadron airman) 31

MacLennan, J.A., Flying Officer (F/O) 133

MacLeod, Don, Corporal (Cpl), (Bandsman) 206, 212

Magee, Pilot Officer (P/O) 36

Malloy, D.G., Flight Lieutenant (F/L), Squadron Leader (S/L) 53, 58, 218

Manns, K., Flying Officer (F/O) 122

Markgraf, Dale, USAF Colonel 175

Markgraf, Dick, USAF Colonel 175

Maridor, Jean-Marie 53

Marrin, Vince, Aircraftman (AC) 11

Martin, Mr. Watt 176

Martyn, M.P., Group Captain (G/C) 114

Mary Otter Trophy 177

Massier, M. "Maryalyce", Captain (Capt), Major (Maj) 171, 175

Maurice, J.E., Flight Lieutenant (F/L) 87, 90

Maville, John (Squadron airman) 31

May, Bob (Squadron airman) 136

McAllister, Dave, Chief Warrant Officer (CWO) 194

McDonald, D.W., Captain (Capt) 175

McGonigle, P. "Pearl", Lieutenant Governor of Manitoba, Honourary Colonel 175, 188, 192

McGraw, G.C., Sergeant (Sgt) 36

McGraw, H.E., Sergeant (Sgt) 58

McGregor, G.R., Squadron Leader (S/L), Wing Commander (W/C) 34, 35, 38, 39, 218

McKinstry, K., Corporal (Cpl) 177

MacLeod, J.A., Flying Officer (F/O) 70

McMillan, Squadron Leader (S/L) 134, 144

McNab, E.A., Flying Officer (F/O) 8, 9

McNaughton, Andrew, Major-General 24, 28 notes

McNorgan, P.D. "Pat", Corporal (Cpl), Sergeant (Sgt) 162 notes, 186, 187, 194, 198

McKnight, W.L., Flight Lieutenant (F/L) 55

McTavish, Greg, Master Corporal (MCpl) 215

Metropolitan Civil Defence Organization 143

Military Cross 107, 170

Miller, Erle S. (Squadron airman) 31

Miller, V.H., Flight Sergeant (FSgt) 57

Minto Armouries 3, 6, 17

Minto, Sergeant (Sgt) 118, 139, 165, 166

Mitchner, J., Squadron Leader (S/L) 65, 66

Mobile Striking Force (MSF) 101

Moncrieff, Ernie, Pilot Officer (P/O), Flying Officer (F/O), Group Captain (G/C) 9, 12, 23, 34, 217

Moore, L.A., Squadron Leader (S/L) 82, 84, 86, 218

Moosomin Flying Club 160

Morris, J.A., Flying Officer (F/O) 66

Morrow, R.E.E. "Bob", Flying Officer (F/O), Flight Lieutenant (F/L), Squadron Leader (S/L) 36-38, 42-45, 47, 51-53, 218

Mulock, R.H. 4

Mulroney, Brian, Prime Minister of Canada 186

Multinational Force and Observers 181

Munn, Maurice (former Squadron member) 123, 162 *notes*

Murchie, G. (Squadron airman) 34

Murphy, N.P. "Spud", Pilot Officer (P/O) 31, 71

Murray, D., Master Corporal (MCpl) 192

Murray, H.L., Flight Lieutenant (F/L) 86

Mynarski, Andrew, Pilot Officer (P/O) 100, 136

NATO 177

Neal, T., Second Lieutenant (2Lt) 192

Nicholson, H.C., Flying Officer (F/O) 81, 84

Nijmegen/Arnhem 76

Non-Permanent Active Air Force (NPAAF) 1, 3, 4, 14, 22

NORAD 161 *notes*, 194, 195, 209

Normandy 67, 70, 90, 95, 193

North Coates Wing 67

North West Air Command 102, 106, 110, 111, 114-116, 119

Northcott, G.W., Squadron Leader (S/L), Wing Commander (W/C) 61, 64-66, 71, 82, 218

Northern Region Headquarters (NRHQ) 172

Nuclear Defence Fallout Recording Sites 156

O'Brian, J.A., Flight Lieutenant (F/L) 89

O'Grady, Mrs. C.E., 151

O'Hagan, W.G., Flying Officer (F/O) 71

O'Leary, J.S. "Jeff", Master Corporal (MCpl) 181, 189

O'Neill, B.P., Sergeant (Sgt), Flight Sergeant (FSgt) 45, 46

Olson, L.E., Major (Maj), Lieutenant-Colonel (LCol) 175, 177

Olson, T., Private (Pte) 177

Oosterbeek War Cemetery, Arnhem, Holland 86

Operation Air Display 126

Operation Airlift 159

Operation Architect 119

Operation Barbarossa 40, 41

Operation Bodenplatte 81

Operation Centre Punch 136

Operation Cerberus 46, 47

Operation Channel Dash 46

Operation Chestnut II 108

Operation Determination 192

Operation Double Exposure 154

Operation Freshie 129, 131

Operation Goldeye II 144

Operation Gunner 126

Operation John 146

INDEX

Operation Jubilee 56

Operation Lucky 126

Operation MacDonald 135

Operation Phoenix 192

Operation Point Blank 63

Operation Portage 111

Operation Santa Claus 158, 160

Operation Sea Lion 32

Operation Sneak Peek 159

Operation Snowshoe 156, 158

Operation Swift 125

Operation Torch 58

Operation Totem II 159

Operation Turnabout 125

Operation Kangaroo 130

Operational Exercise Key Step 126, 127

Operational Exercise Key Stone 126

Osborne, D.L., Flight Lieutenant (F/L) 133

Owen, J.E., Leading Aircraftman (LAC) 38

Osztian, Sean, Private (Pte) 210

Pacholka, W.C., Squadron Leader (S/L) 118, 146

Palton, H.L., Sergeant (Sgt) 36

Patriarche, Pilot Officer (P/O) 6, 217

Patrick, E.I., Brigadier-General (BGen) 179

Patrick, R.W., Lieutenant-Colonel (LCol) 162, 181, 188, 210, 218

Patterson, J.T. "Tom", Squadron Leader (S/L), Wing Commander (W/C) 154, 162, 218

Pawley, Howard, Premier of Manitoba 208

Penner, E., Private (Pte) 192

Pentland, W.H., Pilot Officer (P/O), Flying Officer (F/O) 12, 36

Permanent Active Air Force 22

Permanent Force 3, 7, 16, 17, 24, 110

Peterson, G.F., Pilot Officer (P/O) 86

Pickering, Sergeant (Sgt) 16

Portage la Prairie aerodrome (see RCAF Station Portage la Prairie) 111

Prairie Command (Army) 102

Princess Patricia's Canadian Light Infantry (PPCLI) 16, 110, 177

Proulx, Bennie, Aircraftman (AC) 10

Puff Target 13, 14, 16

   practice 16

   Trainer 25

   training 18

R-boat (also *"Räumboote"*) 63

RAF Beaufighter 63, 67, 69

RAF Station Catterick 66

RAF Station Colerne 50

RAF Station Coltishall 36, 63

RAF Station Digby 33, 66

RAF Station Duxford 41

RAF Station Fairwood Common 50, 52

RAF Station Hawkinge 71

RAF Station High Post 30

RAF Station Horne 68

RAF Station Ibsley 49

RAF Station Kenley 2, 55, 60

RAF Station Martlesham Heath 41

RAF Station Merston 62, 70

RAF Station Peterhead (Scotland) 67

RAF Station Southend/Rochford 41

RAF Station Warmwell 44, 82

RAF Station Wellingore 62, 66

RAF Station Westhampnett 70

Ramsay, D.L., Pilot Officer (P/O) 36

Rang de Fliers 43

Ranville War Cemetery, Calvados, France 62

Ratcliff, A.G., Flying Officer (F/O) 88

Rathwell, D.W., Squadron Leader (S/L), Wing Commander (W/C) 120, 124, 129, 218

Rawlings, P.G., Lieutenant-Colonel (LCol) 190, 218

RCAF (Auxiliary) University Flight 107

RCAF (Reserve) Wing Winnipeg 115, 119

RCAF Station Abbotsford 119

RCAF Station Armstrong 154

RCAF Station Bagotville 128

RCAF Station Chatham 135

RCAF Station Cold Lake 136

RCAF Station Digby 61

RCAF Station Gimli 105, 109, 152

RCAF Station Gypsumville 155

RCAF Station Namao 119

RCAF Station Pagwa 154

RCAF Station Rockliffe 26, 128

RCAF Station Saskatoon 139

RCAF Station Senneterre 141

RCAF Station St. Hubert 130, 134

RCAF Station MacDonald 135

RCAF Station Penhold 142, 157, 160

RCAF Station Portage 133

RCAF Station Trenton 22

RCAF Station Uplands 128

RCAF Station Watson Lake 121

RCAF Station Winisk 148

RCAF Station Winnipeg 8, 15, 97, 104, 111, 121, 136, 146, 150

RF 4402 160, **163** *notes*

Reay, John Jr, Captain (Capt), Pipe Major, Band Officer 199

Reay, John Sr, Flying Officer (F/O) 199, 200

Reeve, J. "Jack", Captain (Capt) (see also Jack Reeve Memorial Award) 171, 173, 184

Reichswald Forest War Cemetery, Kleve, Germany 84

Reid, J.M. "Matt", Flight Lieutenant (F/L), Wing Commander (W/C) 106, 130, 134, 135, 139, 143, **161** *notes*, **162** *notes*, 200, 218

Reid, Stan (Squadron airman) 31

Reserve Pilot Training Program 181

Reserve Tradesmen Training Plan (RTTP) 126, 137

Reyno, E.M., Flight Lieutenant (F/L) 36

Richardson, James A. 4, 5, 13, 14, 20

Riddell, D.B. (Squadron airman) 214

Rigby, J.E., Flying Officer (F/O) 90

Rivers 104, 105, 109, 112, 115, 120, 125, 133, 136, 151, 157

INDEX

Roberts, S., Corporal (Cpl) 85, 191

Robertson, G.D. "Graham", Sergeant (Sgt) 34, 36-38, 46

Roblin, Duff (402 airman) 124

Ross, E.J., Sergeant (Sgt) 36

Royal Canadian Army Medical Corps (RCAMC) 23, 25

Royal Canadian Medical Corps 23

Royal Canadian School of Artilley Shilo (RCSA Shilo) (also called Shilo Range) 105, 108, 121, 125, 126

Royal Flying Corps 6, 32

Runnymede War Memorial, Englefield Green, Egham, Surrey 42, 59, 71

Russell, F., Flying Officer (F/O) 135

Russell, G.A., Pilot Officer (P/O) 34

Russell, H., Pilot Officer (P/O) 36

Salome, R. "Ron", Captain (Capt) 149, 171

SAR Dobbs 151

SAR Green 159

SAR Harrison 146

SAR Totem 150

SAR Macleod 150

Saul, R.E., Air Vice Marshal (A/V/M) 34, 35

Sawchuk, Corporal (Cpl) 130

Schmitt, D.W., Captain (Capt) 175

Schock, S.L., Lieutenant-Colonel (LCol) 197, 218

Schwindt, R.S., Master Corporal (MCpl) 181

Scopwick Church Burial Ground, Lincolnshire, England 38, 66

Scott, Alex (Squadron airman) 31

Scott, D.R., Wing Commander (W/C) 151, 155, 158, 218

Scott, H.N., Flying Officer (F/O), Group Captain (G/C), Honourary Colonel 107, 177, 179, 201

Scott, J.R., Pilot Officer (P/O) 36

Scott, R.A., Captain (Capt) 171, 173

Scott, Warrant Officer (WO) 179

Searles, F.D., Squadron Leader (S/L) 116

Second Tactical Air Force (2nd TAF) 68, 71, 74

Secretan, T. "Bull", Chief Warrant Officer (CWO) 194

Sellers, Flying Officer (F/O) 23, 217

semi-automatic ground environment (SAGE) system 149

Semper, A.W.N., Private (Pte) 192

Sherk, D. Flying Officer (F/O) 66, 80

Short Service Commission (SSSC) 137

Sir James Dunn Award of Excellence 183

Skinner, A.M., Sergeant (Sgt), Pilot Officer (P/O) 36, 59, 60

Slaughter, R.W., Lieutenant-Colonel (LCol) 174, 218

Sleep, K.S., Flight Lieutenant (F/L) 77, 78, 81, 82

Smallwood, Bill (402 airman) 118

Smith, D.J., Pilot Officer (P/O) 36, 44

Smythe, R., Corporal (Cpl) 177

Souvenir Cemetery, Longuenesse, France 58

Speare, A.R., Flight Lieutenant (F/L) 77

Spitfire Lounge 178

Squadron Aircraft Maintenance Engineering Organization (SAMEO) 186, 187, 192, 194, 195

Squadron Family Day 145

SS Duchess of Atholl 28

SS Duchess of York 27

St. Clement Danes Church 180

St. James Legion 177, 215

St. Omer Cemetery, Longuenesse, France 64

St. Pierre. J., Flying Officer (F/O) 38

Sterry, Warrant Officer (WO) 179

Stevenson Field 9, 19, 98, 119, 128

Stevenson, Fred 19

Sully, J.A, Squadron Leader (S/L) 5, 15, 22, 197, 217, 218

Sunstrum, M.J. "Sunny", Pilot Officer (P/O), Flying Officer (F/O) 59, 60

Swanwick, Keith, Captain (Capt) (Air Command Band) 208

Symonds, J.M., Lieutenant-Colonel (LCol) 181, 218

Symonds, James, Corporal (Cpl) 210

Tactical Air Group (see also North West Air Command and No.11 Group) 119, 121, 122, 126, 128, 190

Tactical Group (see also No.11 Group) 108, 115-117

Technical Training Unit (TTU) 115, 116, 153, 185

Teslin Lake Range 123, 124

"The Angel" (See also Chadburn, L.V.) 62

The "Skyhawks" 184

The "Untouchables" 105

Thistle Pipe Band 199, 200

Thompson, J.A., Pilot Officer (P/O), Flying Officer (F/O) 36, 38

Thomson, Ian, Flying Officer (F/O) 129, 133

Thornycroft, Brigadier-General (BGen) 171

Tighe, J.F., Aircraftman First Class (AC1) 38

Toews, Art, Flying Officer (F/O) 133

Torney, J.G., Flying Officer (F/O) 62

Total Force Concept 181

Totem Talk 144, 166

Tottle, Davy (Squadron airman) 27

Tracey, Bob, Major (Maj) 209

Trans-Canada Airlines 19, 118

Trask, N.B., Sergeant (Sgt), Pilot Officer (P/O), Flight Lieutenant (F/L) 36, 44

Treen, G., Corporal (Cpl) 174

Tri-Service Day 131

Tuck, R.S., Wing Commander (W/C) 55

Tweedie, A.S.R., Flight Lieutenant (F/L) 116

United Nations 158, 181, 190

   and India-Pakistan Observer Mission (UNIPOM) 158

University of Manitoba (U of M) Flight 107, 110, 131

Urquart, John, Flight Lieutenant (F/L) 124

V-1 53, 65, 69, 70, 72, 73, 90, 95

V-2 65, 74, 78

Van Helvert, P.J., Corporal (Cpl) 192

Vandahl, E., Flight Sergeant (FSgt) 99

Varey, W., Warrant Officer Second Class (WO2) 150

Vernon Bastable Memorial Award 108

Vickers, A.H., Flying Officer (F/O) 72

Victoria Cross 91

Viens, L.J.J., Warrant Officer (WO) 192

INDEX

Victoria Cross 91

Viens, L.J.J., Warrant Officer (WO) 192

Villeneauve, Bernie 31

Voxair 162 *notes*, 163 *notes*, 178, 198, 210, **216** *notes*

Wagner, E.T., Lieutenant-Colonel (LCol) 172, 218

Waits, Alf (Squadron airman) 8

Waldie, Frank, Flight Sergeant (FSgt), Warrant Officer First Class (WO1) 130 150

Walker, G.W. "Jerry", Sergeant (Sgt) 36, 38

Walker, J.E., Pilot Officer (P/O) 36

Walker, R.R., Pilot Officer (P/O) 36

War Cemetery, Pihen-les-Guines, Pas-de-Calais, France 58

War Cemetery, Grandcourt, France 59

Webb, Mike, Chief Warrant Officer (CWO) 194

Weber, B., Captain (Capt) 171

Weber, Dennis, Chief Warrant Officer (CWO) 181, 194

Wellingore – See RAF Station Wellingore 62, 66

Western Air Command 17, 23, 102

Western Canada Aviation Museum 175, 177, 184

Western Canadian Airways 4, 13

Western Canadian Pipe Band Competition 136

Western Command (Army) 102, 110, 128

White Paper on Defence, 1951-52 120, 121

Whittaker, W.D., Flying Officer (F/O) 73, 77

Willis, Errick, Lieutenant Governor of Manitoba 150

Wilson, A.H., Wing Commander (W/C) 14

Wilson, D.M. "Dave", Corporal (Cpl), Master Corporal (MCpl) 178, 184, 189

Wilson, G.T., Chief Warrant Officer (CWO) 178, 181

Wilson, R.D., Lieutenant-Colonel (LCol) 171, 218

Wing Commander's Commendation 191

Winnipeg Board of Trade 5

Winnipeg Flying Club 6, 145

Winnipeg Light Infantry 23

Witherden, Rick, Lieutenant-Colonel (LCol) 197, 218

Wolkowski, E.R., Flight Lieutenant (F/L) 118, 146

Woodward, F., Captain (Capt) 177

Yeandle, Flying Officer (F/O) 36

Young, W.O., Flying Officer (F/O) 88